The Post-Earthquake City

This book critically assesses Christchurch, New Zealand as an evolving post-earthquake city. It examines the impact of the 2010–13 Canterbury earthquake sequence, employing a chronological structure to consider 'damage and displacement', 'recovery and renewal' and 'the city in transition'.

It offers a framework for understanding the multiple experiences and realities of post-earthquake recovery. It details how the rebuilding of the city has occurred and examines what has arisen in the context of an unprecedented opportunity to refashion land uses and social experience from the ground up. A recurring tension is observed between the desire and tendency of some to reproduce previous urban orthodoxies and the experimental efforts of others to fashion new cultures of progressive place-making and attention to the more-than-human city. The book offers several lessons for understanding disaster recovery in cities. It illuminates the opportunities disasters create for both the reassertion of the familiar and the emergence of the new; highlights the divergence of lived experience during recovery; and considers the extent to which a post-disaster city is prepared for likely climate futures.

The book will be valuable reading for critical disaster researchers as well as geographers, sociologists, urban planners and policy makers interested in disaster recovery.

Paul Cloke was Emeritus Professor of Human Geography, University of Exeter.

David Conradson is Professor of Human Geography, University of Canterbury.

Eric Pawson is Emeritus Professor of Geography, University of Canterbury.

Harvey C. Perkins is Emeritus Professor of Planning, University of Auckland.

Routledge Studies in Hazards, Disaster Risk and Climate Change

Series Editor: Ilan Kelman, Professor of Disasters and Health at the Institute for Risk and Disaster Reduction (IRDR) and the Institute for Global Health (IGH), University College London (UCL)

This series provides a forum for original and vibrant research. It offers contributions from each of these communities as well as innovative titles that examine the links between hazards, disasters and climate change, to bring these schools of thought closer together. This series promotes interdisciplinary scholarly work that is empirically and theoretically informed, with titles reflecting the wealth of research being undertaken in these diverse and exciting fields.

Empowerment and Social Justice in the Wake of Disasters
Occupy Sandy in Rockaway after Hurricane Sandy, USA
Sara Bondesson

Climate Change and Risk in South and Southeast Asia
Sociopolitical Perspectives
Edited by Devendraraj Madhanagopal and Salim Momtaz

Gender-Based Violence and Layered Disasters
Place, Culture and Survival
Nahid Rezwana and Rachel Pain

The Post-Earthquake City
Disaster and Recovery in Christchurch, New Zealand
Paul Cloke, David Conradson, Eric Pawson and Harvey C. Perkins

Local Adaptation to Climate Change in South India
Challenges and the Future in the Tsunami-hit Coastal Regions
Devendraraj Madhanagopal

For more information about this series, please visit: https://www.routledge. com/Routledge-Studies-in-Hazards-Disaster-Risk-and-Climate-Change/ book-series/HDC

The Post-Earthquake City

Disaster and Recovery in Christchurch, New Zealand

**Paul Cloke, David Conradson,
Eric Pawson and Harvey C. Perkins**

Routledge
Taylor & Francis Group

LONDON AND NEW YORK

First published 2023
by Routledge
4 Park Square, Milton Park, Abingdon, Oxon OX14 4RN

and by Routledge
605 Third Avenue, New York, NY 10158

Routledge is an imprint of the Taylor & Francis Group, an informa business

British Library Cataloguing-in-Publication Data
A catalogue record for this book is available from the British Library

ISBN: 978-0-367-22552-0 (hbk)
ISBN: 978-1-032-43672-2 (pbk)
ISBN: 978-0-429-27556-2 (ebk)

DOI: 10.4324/9780429275562

Typeset in Times New Roman
by SPi Technologies India Pvt Ltd (Straive)

Contents

Figures

Tables

Preface

At the time, the Canterbury earthquakes garnered a lot of international media coverage, but as with all such disasters, wider interest quickly faded. One of the stranger things about this event, however, is that it grew into a series, having begun without warning in early September 2010 and continuing, not with a diminishing number of aftershocks as was predicted, but with renewed energy and damaging major shakes over a period of some years. The most destructive one was in late February 2011, and it generated one of the largest-ever claims on the international insurance industry, despite being of much lesser magnitude than many 'regular' earthquakes elsewhere.

The Canterbury earthquake story is marked by such peculiarities, which in our view have long warranted both attention and explanation. But although there has been a lot of work on aspects of the events published in the periodical literature, official sources and the media, there has not before now been a coherently argued account from a social science perspective. A recent set of essays (Uekusa et al. 2022) is an exception and is valuable in recognising that the city of Christchurch at the centre of our story has become something of a talisman for the globally emerging experience of multi-hazards. The earthquakes have been followed by floods and fires, the mosque shootings of March 2019, which once again brought the attention of the world's media, and most recently Covid-19. The Covid lockdowns have in turn cast doubt on some of the assumptions on which the post-earthquake rebuild has been based. Yet all along, those assumptions – of government intervention and showcase projects – have been complemented and countered by grassroots initiatives and transitional experiments. The shape and outcomes of the 'rebuild' have therefore played out in unpredictable ways of a period of years.

This is one reason why we have waited more than a decade to produce this book, and why we have taken our time to engage with a still evolving process, and with each other as authors. We know Christchurch and Canterbury well, three as long-term residents who lived through the earthquake sequence, and the fourth as a regular visitor to the city over many years, both before and since the event. We provide outline biographies on page xiv and discuss our various positionalities in the first chapter.

We owe debts of gratitude to a large number of people, starting with Tim Nolan, our talented and ever patient cartographer. Amongst many who have helped, encouraged or influenced us are Thomas Blakie, Jacky Bowring, Jan Cameron, Kelli Campbell, Ikenna Chukwudumogu, Simon Dickinson, Kelly Dombroski, Erin Heine, Alan Latham, Deborah Levy, Robin Kearns, Rob Kerr, Simon Kingham, John McDonagh, Mike Mackay, Colin Meurk, Judi Miller, Garry Moore and members of the Tuesday Club, Laurence Murphy, Brent Nahkies, Hugh Nicholson, Mark Poskitt, Nick Taylor, Sarah Tucker and Alice Ann Wetzel. We also thank all those who agreed to be interviewed for our study and the Christchurch journalists whose comprehensive earthquake recovery reporting has been invaluable. We also acknowledge the individuals and agencies who have allowed us to reproduce images; they are attributed in the usual way in the captions to the figures within each chapter. All unattributed graphics or images have been sourced from within the author team.

Sadly, Paul Cloke died suddenly when the book was in the final stages of writing. He wished to acknowledge funding from the Erskine programme at the University of Canterbury and the Leverhulme Trust Emeritus Fellowship programme, which enabled his travel over the years to Christchurch, as well as the International Office of the University of Exeter. The book was originally Paul's idea, and it has come to fruition as a result of his inspiration and encouragement. We hope that he would have been proud of the final result.

Christchurch, New Zealand
June 2022

About the authors

Paul Cloke was Emeritus Professor of Human Geography at the University of Exeter. His research interests included rural geography, social change, ethical geographies and the role of the third sector. Paul published a number of papers with New Zealand colleagues (including each of us) and co-authored a number of academic articles on aspects of the aftermath of the Christchurch earthquakes. During his career, he produced over 40 books, including the co-authored *Geographies of Postsecularity: Re-envisioning Politics, Subjectivity and Ethics* (2019). In 2022, Paul was awarded the Royal Geographical Society's Victoria Medal for his contribution to rural geography and to the wider discipline.

David Conradson is Professor of Human Geography at the University of Canterbury. His current research examines the lived experience of disrupted environments, with a focus on processes that shape individual and collective well-being. He has been involved in a number of funded research projects examining post-disaster recovery in Christchurch, with publications from this work in *Transactions of the Institute of British Geographers, the International Journal of Disaster Risk Reduction, Earthquake Spectra*, and a series of book chapters. Previously an editor of *Social and Cultural Geography*, he is currently the Managing Editor of the *New Zealand Geographer*.

Eric Pawson is Emeritus Professor of Geography at the University of Canterbury. He retains an active interest in research, contributing to the *Cambridge Handbook of Undergraduate Research* (2022) and the fifth edition of *Qualitative Research in Human Geography* (2021). His thematic interests are in environmental history and environmental governance, with publications including *The New Biological Economy* (2018). He has been involved in a range of organisations and initiatives in post-earthquake Christchurch, particularly in and about the residential red zone. He co-founded the Ōtākaro Living Laboratory Trust and has chaired the Waitākiri Ecosanctuary Trust.

Harvey C. Perkins is Emeritus Professor of Planning at the University of Auckland where he directed the research programme Transforming Cities: Initiatives for Sustainable Futures. He is a Past President of the New Zealand Geographical Society and former Professor of Human Geography at Lincoln University in Christchurch. His research interests include leisure and tourism, housing, residential intensification and the relationships between house and home. He is a Principal Investigator in the New Zealand National Science Challenge *Building Better Homes, Towns and Cities: Ko Ngā Wā Kāinga Hei Whakamāhorahora*, for which he co-leads a study of local regeneration initiatives in regional settlements.

1 Introduction

The years 2010 and 2011 saw major environmental disasters in different parts of the world. Much of the Caribbean state of Haiti was devastated in a magnitude 7.0 earthquake in January 2010, with a death toll in the hundreds of thousands. The next month, 12,000 people lost their lives after an 8.8 magnitude quake and tsunami in Chile. In March 2011, 6,000 people were killed in the magnitude 9.1 shock and tsunami at Tōhuku in Honshu, Japan. In between these events, a long-running seismic sequence began in the vicinity of the south Pacific city of Christchurch, in Aotearoa New Zealand's South Island. Starting on 4 September 2010, the Canterbury earthquake sequence (named for the region in which the city is located) continued for more than three years, including a very damaging, shallow 6.3 magnitude event under the city on 22 February 2011. The Canterbury earthquakes were not as severe as the other events, nor were they as deadly, although the official earthquake toll was 185 people (New Zealand Police 2022).

Nonetheless, this particular event quickly became one of the world's most expensive environmental disasters. As Swiss Re, the international re-insurance company, asked, how 'could a small aftershock in a city not considered an earthquake hotspot trigger one of the largest insurance losses ever?' (Grollimund 2014). And what did this financially framed question mean in terms of human experience and response? Some aspects of the recovery process attracted wide notice internationally, such as the Student Volunteer Army, the transitional city movement spearheaded by Gap Filler, the downtown container mall and street art festivals. But what of more quotidian encounters and experiences in a place where over a decade later, the process of recovery still continues? Scars remain on people and homes, even if superficially damage outside the city centre is now less visible. In the centre itself, where most buildings were lost, large empty sites remain to be redeveloped, whereas extensive greenfield areas of new housing now fringe the metropolitan area.

This book presents a critical assessment of Christchurch as an evolving post-disaster city, recognising that such a narrative is both intensely local in scale and also of global significance. From the neoliberal economic experimentation and agricultural deregulation of the 1980s (Le Heron and Pawson 1996) to the hospitable welcome of multicultural immigration and the one-nation

DOI: 10.4324/9780429275562-1

revulsion of terror in the 2010s, and the fighting of Covid-19 with a team of five million in the 2020s, New Zealand has become a place of interest to many beyond its shores (Uekusa et al. 2022). The rebuilding of Christchurch has occurred as a result of very specific local and national interventions and community-based activities, yet the city serves as a wider laboratory in which a tragic period of destruction, disruption and disaster has in turn been followed by rare opportunities to refashion land uses and social experience from the ground up. Evaluation of what has been possible in these enforced laboratory conditions offers wider international lessons regarding both the powerful magnetism of reproducing previous normalities and the capacity to integrate experimental cultures of place-making and ecological restoration into the urban fabric.

This introductory chapter addresses the contexts in which the narratives of the book are embedded and identifies some of the key points that need to be interrogated. First, we outline some of the historical, environmental and political particularities of Christchurch's situation, as the characteristics of the pre-disaster city have certainly shaped events after the earthquakes. Second, we draw on existing wisdom in disaster models and narratives in the literature, outlining (sometimes potentially conflicting) key theoretical ideas which help to shape the different strands of our account of the post-quake experience. Third, we acknowledge that the ways in which these matters are framed and narrated reflect the complex positional and situated knowledges of the authors. This not only recognises the existence of our different narrative perspectives but also the ways such perspectives coexist in a multifaceted web of understanding.

Contextualising Ōtautahi Christchurch

There was no physical connection between the earthquakes in different countries in 2010–11, but it is no coincidence that Chile, Japan and New Zealand lie on the Pacific Ring of Fire, the active zone of volcanism and seismicity that marks tectonic plate boundaries in the western hemisphere. However, as Swiss Re observed, it was always expected that in New Zealand the capital city of Wellington would be at greatest risk, as it sits astride the fault lines that mark the boundary between the Indian and Pacific plates (Grollimund 2014). Yet it has long been recognised that there are few places in the country that are entirely free of earthquake hazard. 'Living in New Zealand means living with earthquakes' (Department of Statistics 2010, 6) and under amendments to the Building Act 2004, Christchurch is in the highest risk category. By the 1990s, the likelihood of liquefaction, or land deformation under seismic stress, was known, an environmental vulnerability that the city shares with many other Pacific ring cities, such as Jakarta, Tokyo and Vancouver. But why then are such prominent places built in sites of high risk?

An explanation lies in what the novelist Amitav Ghosh has recently called 'terraforming'. He borrows this concept from science fiction to describe the process of land-making or 'land-moulding' that typified colonial settler experience, 'envisioning space as a "frontier" to be "conquered" and "colonized"' (Ghosh 2021, 54). Such colonising, replicated on steroids during the urbanisation associated with the current era of neoliberal globalisation, treats the environment, itself an objectifying concept, as an inert stage on which humans as active agents inscribe their 'improvements' (Latour 2018). Christchurch, the centre of the planned Victorian colony of Canterbury, was planted in 1850 in the midst of swampy and sandy ground, without awareness that this is relatively recently formed land at the edge of the delta of the Waimakariri River, one of the great, unpredictable gravel-bed rivers that flow eastwards to the Pacific Ocean from the Southern Alps (Figure 1.1). Whilst much of the history of the city has been preoccupied with land drainage and exotic revegetation (Schrader 1997; Pawson 2000), the hidden price of this has been development on deltaic sediments prone to liquefaction during earthquakes, as Chapters 2 and 3 illustrate in greater detail.

Another way to describe this is to suggest that the land is 'alive' in its own right, a perspective embodied in the worldview of the indigenous Māori, who were largely displaced by nineteenth-century colonisation. They were however by no means silenced. One of the key features of contemporary governance in New Zealand has been the growing recognition, both politically and in legislation, of the cultural rights of Māori iwi and hapū (tribes and subtribes), stemming from the signing of the Treaty of Waitangi between the British Crown and iwi in 1840 (Orange 1987; Finlayson and Christmas 2021). Settlements to rectify breaches of the Treaty have been ongoing since the 1990s, one of the earliest and biggest being with the South Island iwi, Ngāi Tahu, in 1998. Canterbury lies within the rohe, or area within which Ngāi Tahu has mana whenua, or cultural and customary authority (Tau 2016). For this reason, the term 'the Crown' is in regular usage in New Zealand to denote central government in its exercise of the authority originally assumed through the Treaty by the colonising power. And in post-earthquake legislation, Ngāi Tahu has been identified as a partner with central and local governments, and state agencies, in the rebuild of the city and metropolitan area. The local hapu is Ngāi Tūāhuriri, which exercises mana whenua over much of the area in which the city sits. Māori terms are increasing in everyday usage, including Ōtautahi as the name of the city (Uekusa et al. 2022).

The city is broadly typical of those planted by white colonial settlement worldwide. It developed on a regular grid, quickly outgrowing this along late nineteenth-century tramlines. After World War Two, it expanded outwards under the imperatives of suburbanisation and automobility during a period of national prosperity (Forer 1978), checked only by the barrier of the Port Hills to the south and the river estuaries to the east (Figure 1.1). As Chapter 3 demonstrates, its spread was not socially uniform: suburbs in the northwest and on the hills tend to be more socially privileged than those in the east, and characterised by a denser 'urban forest' of exotic trees. The city has long had a reputation as class divided, with conspicuous traditions energised by the Anglican roots of the colonising organisation and reflected in its Victorian Gothic architecture and network of private schools. But this is at best a

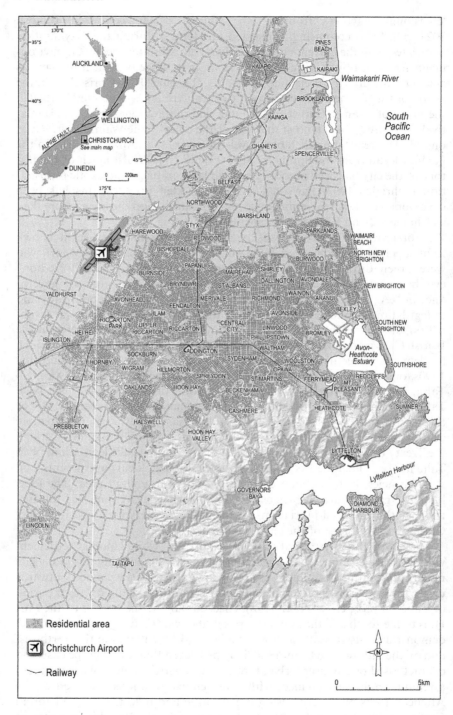

Figure 1.1 Location map.

partial representation of a place that has also exhibited a long-run working class and radical spirit (McAloon 2000a, 2000b). Most of its parliamentary seats are usually held by the left-of-centre Labour Party. Such tensions were for long hidden beneath a popular identity as the 'garden city' of the South Island (Galloway 2022; Pawson 1999).

Theorising disaster-affected places

Within the interdisciplinary disaster and critical disaster studies literature, the process of disaster recovery is often presented as a sequence of stages through which an affected community is expected to progress (eg. Cutter et al. 2008; Kates et al. 2006). Figure 1.2, drawn from DeWolfe's (2000) research in the United States, is illustrative. Although such models provide a useful guide to the stages of recovery, they imply a linear and unified experience, which seldom occurs as depicted. In Christchurch, the pace of recovery varied significantly among households, communities and neighbourhoods across the city, even in adjacent houses. Recovery was also by no means assured, contrary to the optimism inherent in the steady upward movement of DeWolfe's line and its final stage of 'new beginning'. This presents a challenge to recovery models that champion stoicism and assume population resilience. In many disasters, some people and communities are affected very badly, suffering devastating emotional and material losses. It is important that these experiences are not ignored or eclipsed by buoyant models that internalise optimistic progress.

While acknowledging a broad sequence of events in the case of Christchurch, this book gives attention to the multiple co-existing temporalities of disaster experience. This includes noticing the ways in which time may stretch or dilate for those waiting for emergency relief and assistance, as in

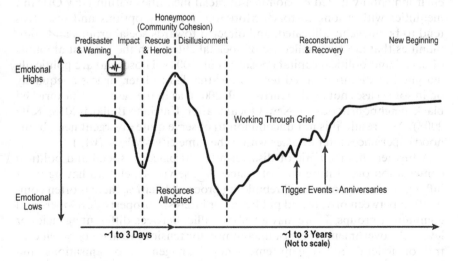

Figure 1.2 DeWolfe's model of post-disaster recovery.

the immediate aftermath of a major earthquake when buildings may be shattered and communication networks disrupted. Other people may experience time as somehow suspended, as they navigate complex insurance claims and await government decisions amidst repeated bureaucratic delays and impasses. For still others, the sudden violence of post-disaster life is foregrounded, such as when people are killed or injured, or when houses and buildings are demolished, erasing once familiar landmarks and places of belonging.

Disaster-affected places have been understood as having certain characteristics. Evidence from the Haiti earthquakes of 2010 and 2021 (DesRoches et al. 2011; Hou and Shi 2011; International Labour Organisation 2021) emphasises disruption to the connections and relationships that sustain and enable the reproduction of places and homes. Damage to transport infrastructure upsets the usual functioning of places, compounding shock, deprivation and hardship. But places are also pulled into new networks of relationships, as sites of humanitarian relief, charitable donations, government and insurance resources and media attention. Opportunities may also beckon for opportunistic profiteering by agents of disaster capitalism (Klein 2008; Miles 2016). Organisational disorientation amongst government and community agencies (Hall et al. 2016) is mirrored at individual and household levels, where disruption can be intensified by the severance of local place attachments (Fullilove 2016).

Another significant characteristic of disaster-affected places is the exposure and often exacerbation of pre-existing inequalities. The widely observed stratification of vulnerability and disadvantage by socio-economic status (Kahn 2005), race (Cutter et al. 2006), gender (Enarson and Pease 2016) and disability and sexual orientation (Dominey-Howes et al. 2013; Gaillard et al. 2017) is typically highlighted during disasters (Wisner and Luce 1993). As Johnson (2006, 148) argued after Hurricane Katrina, 'the storm laid bare the environmentally linked economic and racial inequities within New Orleans, inequities with a long history'. Moreover, recovery options and resources tend to be unevenly allocated, and directed towards social groups and communities that are most reflective of societal elites with the greatest amounts of social and political capital (Wisner et al. 2004). Those who are politically marginal or disenfranchised tend to be attended to later, or less adequately or, in some cases not at all (Curtis et al. 2007), as in the case of impoverished black neighbourhoods in New Orleans after Katrina (Bullard 2018; Katz 2008). As a result, it is not uncommon to observe quite divergent neighbourhood experiences after disaster within the same city (Wilson 2013).

A further characteristic of disaster-affected places is social and political contestation over future uses of space. Whose visions end up having most influence in the recovery and rebuilding process? In cities, there is often competition between private and public sector interests, property developers and community groups. There may also be conflict between different agencies or levels of government. It is not uncommon for tensions to arise between central or federal government emergency management organisations and locally elected city and metropolitan authorities, where the former tend towards rapid and relatively undifferentiated responses that are implemented at scale and the latter are, by nature, more attuned to local differences in community and neighbourhood constitution and needs (Hörhager 2015; Pelling 2010). Indeed, much conceptualisation of post-disaster cities echoes the sentiment expressed by Roberta Brandes Gratz (2015, 128) in the context of post-Katrina New Orleans, that 'considerably more wisdom and careful urban regeneration have been exhibited from the bottom up than from the top down'.

Despite their underlying complexity and context dependency, these broad disaster characteristics are often further generalised into ideas about resilience. This has become a pervasive concept in disaster studies over the last 20 years (Soens 2020). According to Adger (2000), resilience reflects the capacity of groups and communities to cope with the disruptions and disturbances of social, political and environmental change. In these terms, resilience is deployed as an attractive yardstick of the ability of a population to accommodate shocks and adapt (Parker 2020) and to 'bounce back' (Manyena et al. 2011). There are, however, a number of significant limitations to such deployment. At a basic conceptual level, resilience is often ill-defined in relation to other elements of community reaction to change, especially that of resistance.

Wenger (2017) questions whether resilience resembles the oak (with its capacity for resistance, enabling absorption of stress and return to a stable equilibrium) or the reed (with its pliability and flexibility to move with changing conditions) or indeed both. Attention to the post-disaster city would seem to require a conceptual framework that acknowledges a co-existence of contemporaneous capacities for both continuity and change. In addition, in assessing the conceptual power of resilience, it is important to consider Hayward's (2013) argument that its usage tends to be conservative. She suggests that it privileges the social constructs of external experts (such as urban planners) and may underestimate the potential for different forms of political resistance that unleash alternative imaginations of new beginnings and different outcomes. At worst, undue attention to resilience within neoliberalised ways of thinking can lead to an attribution of responsibility for survival and change to individuals rather than to shared societal structures (Matthewman and Uekusa 2021).

Accordingly, this book offers a critique of what might be described as 'the orthodoxy of resilience' within the disaster recovery literature. Although the capacities and creativity of the individuals, households and communities affected by the earthquakes are acknowledged, these capacities are not enrolled into a tidy narrative of resilience, stoicism and coping which elides the complexity of post-disaster life and experience. There are plenty of instances when the experience of residents and communities in Christchurch defy such categories, or certainly trouble their inherent optimism. Following Donna Haraway (2016), albeit in a different register than originally intended, this book therefore seeks to 'stay with the trouble' rather than move too quickly to optimistic notions of progress, success, resilience, fortitude and 'moving forward'.

In place of these kinds of grand narratives of resilience, the book examines how capacities for continuity, experimentation and innovation are interwoven in the complex spaces, political cultures and ethical values of a post-disaster city. Following the arguments outlined by Cloke et al. (2017), it frames the Canterbury earthquakes in terms of what the French philosopher Alain Badiou (2005) terms an 'event'. Badiou argues that a major event is likely to bring about social change based on two overlapping realms of human action. The first is characterised by an instinct to act in what might be considered ordinary and familiar ways, in line with and managed by dominant ideologies and their consequent influence on the activities of the state. The expectation here is that pre-event norms will re-emerge, as actors seek a return to what they know. The second possible outcome is the emergence of alternative possibilities for action that had previously been submerged, covered up or excluded. In these terms, the significance of the event is that it has the capacity to rupture the established order of things and to allow this second strand of action to take flight, affording a distinctive space that has the potential to open up new beginnings and new imaginations of collective cultural creativity (Bassett 2008; Dewsbury 2007).

In strict terms, the Christchurch disaster needs to be acknowledged as a *sequence* of geological events, with the two destructive earthquakes accompanied by over 11,000 aftershocks of varying magnitudes, as illustrated in Chapter 2. In many respects, then, it differs from Badiou's emphasis on the kind of event brought about by political upheaval. However, geographers have also recognised the power of what they term the 'geo-event' (Shaw 2012; Yusoff 2013) to cause a disruption of inhuman nature (Clark 2011) that in turn can rupture the foundations, structures and assumed relations that make the ordinary and familiar world legible. Thus, it can be argued that the collective outcome of a geo-event such as the sequence of earthquakes in and around Christchurch can also be to tear the fabric of normal practice in the world, and to foster new senses of life and place as well as provoking a desire to return to the status quo.

This notion of the rupture of the event is central to the deliberations in this book, which address the questions of what exactly was ruptured by the 2010–13 Canterbury earthquakes (Pickles 2016), and to what degree there are still spaces for alternative community action in the city a decade later. Subsequent chapters will variously explore capacities for continuity and change, often occurring simultaneously in time and space although at other times significantly variegated in their geographical patterns. In so doing a range of concepts are drawn on to inform the processes by which Christchurch residents have shaped their lives and found their lives shaped. Political economic models attentive to the dynamics of capital and investment help to explain how both city centre redevelopment and suburban sprawl exhibit pathways of continuity as part of a return to normal after the earthquakes. At the same time, cultural and political ideas are used to detect new and experimental forms of interaction within the city, charting an affective politics that is capable of creating new ways of being in common (Cloke and Conradson 2018).

Engagement with more-than-human ecologies and polities also indicates very significant capacities for change. These ideas are far from abstract: they are negotiated and grounded in changing forms of governance in the city, which inevitably becomes a key focus in critical discussions of rupture in post-disaster Christchurch (Pawson 2022).

Positioning the book

The complex and sometimes divergent narratives presented in this book are not only shaped by different theoretically informed perspectives. It has now become accepted practice for researchers in human geography and cognate disciplines to be reflexively conscious of how particular aspects of self-identity help to shape the conduct and reporting of research (Rose 1997). England (1994) famously argued that a researcher's positionality and biography directly affect the dialogical process of research, and that social identifiers such as gender, ethnicity and age often shape the maps of consciousness that guide research and writing. To explore these positional specificities via reflexive practice is to acknowledge that researchers know the world partially, influenced by how their knowledge is socially, spatially and intellectually situated, and that researchers are susceptible to the promotion or inhibition of particular kinds of insights. Such reflexivity risks introspection but nonetheless the four authors of this book consider it relevant and important to recognise some of the ways in which their experiences and intellectual fascinations have influenced the analytical narratives that have emerged.

Three vectors of positionality are of particular significance. First, the authors have different connections and relationships to Christchurch as a place. Three of us call the city home, and have lived through the earthquakes and their aftermath; the fourth was a frequent visitor. Insider knowledge may have positive benefits for research, including a sensitivity to local contexts and a degree of involvement and familiarity that enhances depth of understanding. Whilst outsider knowledge may be less attuned to local conditions, it can provide a bulwark against overfamiliarity by enabling a more dispassionate questioning of locally socialised assumptions and perspectives (Dwyer and Buckle 2009). However, it is also important to recognise the fluidity and intersection between 'insider' and 'outsider' positions. Thus Paul's 'outsider' position as a British-based researcher was tempered by the experience of a dozen or so research trips to the city over a 30-year period, the last being in 2020. The other authors also have quite different experiences as residents of different parts of the city and participants in post-earthquake recovery. As a team, we have drawn on these distinct and varied knowledges of Christchurch in exploring the drivers and consequences of processes of recovery.

Secondly, we carry with us significant intellectual and academic proclivities. Harvey's early research interests in urban growth management and latterly in aspects of urban planning, design and policy, have led to a significant focus on the role of property owners and businesses as well as residents in the processes and practices of urban regeneration. Eric's enduring interests in environmental history and management have been coupled with involvement in post-disaster governance at a range of scales in the city, in community groups and in the future of its red zone. David's long-standing concern with the experience of places disrupted by neoliberal economic restructuring and welfare reform has highlighted the need to understand the social and psychological impact of the earthquakes, and the role of the community and voluntary sector in recovery. Paul's focus on the 'messy middle' ground between how people shape their lives, and how they are shaped by other factors, emphasises the possibility of political and ethical hopefulness in amongst the often morose analyses of neoliberalised political economy. These different priorities help to explain our readiness to explore a range of different aspects of the post-disaster city.

Thirdly, the authors have had varied participatory experience in terms of research and policy-making in post-quake Christchurch. Eric's city-as-laboratory approach has led him to build a city-based network of contacts through participation in community groups and a weekly critical discussion group (the Tuesday Club) for all things post-earthquake. He was contracted for a time to Regenerate Christchurch (the Crown/council rebuild agency) and (independent of that) initiated a city-wide process to explore governance options for the red zone. Before the earthquakes Harvey's research in the city was on tourism development, residential intensification and urban planning, and afterwards he engaged with resident and professional groups about post-earthquake housing, and with the responses of central city property owners to the post-quake environment. David's post-quake research has centred on the fate of displaced red-zoned households, disruptions to schooling, youth well-being, and the capacity of voluntary sector organisations to assist affected people and places. Paul's participation took the form of collaborative research into the role of the voluntary sector in post-quake transition and care, involving interviews with more than 50 stakeholders during three periods of research at different times after the earthquakes.

These various arenas of participation contribute significantly to the architecture of the book, providing a range of different points of entry. It is our conviction that there are many different elements that coalesce in the evolution of the post-disaster city, hence our concern to emphasise complexity, diversity and plurality in analysis of patterns of change, innovation and meaningfulness. If different parts of the book do not always easily cohere, this is a symptom of the 'messy' geographies at work. We have come to distrust singular meta-narratives and headline conclusions. Instead, our preference has been to acknowledge how variations in authorial positionality have enabled emphasis on the multiple epicentres, concerns, activities and voices that need to be recognised as integral to the emergence of Christchurch from the earthquakes. So, we champion small-scale voluntary and community actors as well as the property owners, policy-makers and planners, the creative innovators and the proponents of a return to 'normal', the disadvantaged and uprooted as well as the suburban survivors, and the ecology of the more-than-human as well as anthropocentric orthodoxy. In so doing the book aims to provide a framework in which to assemble the multiple experiences and realities of recovery.

Structure of the book

Following this introduction, the book is structured into three substantive sections reflecting broad sequences of events in post-disaster experience. The first, labelled 'Damage and displacement', focuses on the material damage of the earthquakes and their impact on households and communities in terms of lived experience. Chapter 2 explores why the city was so vulnerable to the earthquakes despite widespread use of measures of protection and security (specifically building codes and the extent of property insurance) to protect against uncertainty. Having considered the material damage, Chapter 3 then examines its impact on people's lived experience. This includes how post-quake land zoning decisions and insurance wrangles created challenges for many households, with associated reductions in well-being and the increased incidence of anxiety and depression. But the novel expressions of support and mutual aid that arose in communities and neighbourhoods, and their contribution to local resilience and recovery, are also considered.

The second section, 'Recovery and renewal', begins with Chapter 4 on governance and cartographies of recovery. This explores pre-earthquake experiments in governance, when collaborative and community-informed processes had been introduced to map out possible city futures, as a basis for understanding reaction to the state-driven recovery mechanisms put in place in 2011, and the panoptical cartographies employed as part of redrawing of the central city in particular. Chapter 5 considers damage to housing and the subsequent housing recovery effort, producing an array of outcomes in the central city, suburbs and satellite towns. The recovery has encouraged residential redevelopment in the central city and parts of the suburbs, but also significantly increased a long-established tendency to urban sprawl. In an illustration of path dependency, recovery has disrupted earlier plans to contain sprawl through intensification of the existing urban fabric.

Chapter 6 discusses how Christchurch's commercial property sector was affected by the earthquake sequence and the building demolition that followed. But it reinforced prior strongly articulated public and private sector commitments to the central city and refocused pre-earthquake attempts to stem decline in the face of growing suburban competition from retailing, hospitality and allied services. The objective of the chapter is to outline how in the recovery process, cooperative, contested and sometimes antagonistic interaction between the city council, Crown agencies, property owners, their insurance companies and the region's residents have led to a partial recovery in a situation of 'staggering complexity'. Chapter 7 focuses on the role of voluntary and community organisations, including those that are faith-based, resident-led and the novel movement that became known as the Student Volunteer Army. It notes the sometimes uneasy coexistence between such voluntarism and the efforts of state recovery agencies, and considers how this voluntarism sits alongside politically conservative aspirations for rebuilding the city on the one hand and more progressive forms of place-making on the other.

The third section, 'the City in Transition', focuses on the emergence of post-recovery futures. Chapter 8 discusses a phenomenon that has received perhaps greater acclaim internationally than locally: how a number of voluntary sector and community-based organisations sprang up to make energetic and multifaceted contributions to cultural, artistic, creative and ecological uses of temporary spaces. In a series of experimental projects, these organisations have presented propositions for alternative expressions of transition after the earthquakes, valorising collective rather than individual ethics, and creative aesthetics that turn flux into opportunity. In contrast, Chapter 9 conceptualises the city as a locus of consumption flows, with commercial forms of place-making that range from those that have borrowed from the experiments of the transitional movement, to major Crown-sponsored anchor projects purposed to support recreation, events and tourism. This spatial competition to attract consumers and consumption practices however depends on and drives a further cycle of consumption, that of carbon.

Spatial unevenness has also been a theme of recovery, and it was not long before the term 'forgotten east' began to circulate, capturing the view that eastern areas of the city were somehow being overlooked or under-prioritised. Chapter 10 interrogates this idea, examining the differing levels of attention displayed by state and private sector agencies in their engagements with communities in the eastern suburbs. Chapter 11 also takes an important theme in urban geography but one often downplayed in post-earthquake recovery, that of the more-than-human city. The chapter explores how actual worlds become assembled when all manner of things (or 'actants') contribute to the making of particular geographies, assessing the extent to which opportunities for more lively spaces, hybrid ecosystems and species interactions are being recognised.

The question arising here is if through its post-earthquake experience the city is now better placed to move beyond what Donna Haraway (2016, 35) has described as the 'unprecedented looking away' of the Anthropocene. One site in which this is being actively played out is the city's residential red zone in the eastern suburbs, abandoned in the wake of the earthquakes. Chapter 12 asks if this will become the city's field 'of dreams', countering a recurrent criticism of the post-earthquake rebuild, that it is producing 'the city of yesterday' rather than that of tomorrow. After exploring processes of community engagement, regeneration planning and environmental co-governance with Māori, it concludes by outlining some of the nascent characteristics of what a future Christchurch might look like, based on this post-earthquake experimentation over the last decade. These themes are considered further in the concluding chapter.

Part I

Damage and displacement

2 The fracturing of a vulnerable city

Introduction

It was a dramatic and seemingly endless experience living through the Canterbury earthquake sequence over the three or more years from when it began, so unexpectedly, in the early hours of 4 September 2010. This chapter assesses the extent and nature of the physical fracturing to the city and its region generated by those events, seeking to place experience in a wider descriptive and analytical frame. It explores the forms that material damage took, and considers the extent to which this was the product of, or exacerbated by, particular vulnerabilities that – with the benefit of hindsight – can be attributed to characteristics of the city's built environment or its social organisation. Given that Christchurch is the second largest city in a country well-known for seismic activity, one that sits astride a major global tectonic plate boundary, why was it not better prepared? Or was it prepared, but in unhelpful or insufficient ways? And can such issues be illuminated by examining them at different spatial scales, from the local to the global?

The chapter starts with a discussion of the concepts of vulnerability, resilience and security. The last of these is particularly important in a country in which an interventionist state historically sought to provide a level of protection and support to its citizens in the face of uncertainty. The technocratic forms assumed by such protection, particularly building codes and the extent of property insurance, are then examined. This helps to answer Swiss Re's question (Grollimund 2014), posed at the start of Chapter 1: why did the Canterbury earthquake sequence become one of the most expensive episodes ever faced by the international insurance industry? Yet, by global standards, none of the earthquakes was particularly large. Resolution of this apparent paradox is sought through examination of the ecology of the earthquakes, in order to understand the intersection of physical and social circumstances which rendered the city particularly vulnerable. That vulnerability is then illustrated with a description of the extent of death and destruction, followed by an outline of the immediate disaster response, so as to provide a clear context for what might be termed a 'socio-seismic' series of occurrences.

DOI: 10.4324/9780429275562-3

Vulnerability, resilience and security

Vulnerability has been a popular term in environmental hazard and disaster studies for 30 years or more, yet is used in a range of different ways (Wisner 2020). In relation to earthquakes, it has often been reduced to assessments of exposure to risk. But although the known incidence of hazardous events is important, of equal significance are the human attributes of particular communities in terms of 'their capacity to anticipate, cope with, resist and recover from' disastrous circumstances (Blaikie et al. 1994, 9). But even this broader definition implies a relatively static understanding of complex social situations. What needs to be anticipated might not be known, and many factors, apart from the characteristics of the event and people affected, shape the nature of impacts. Among these are the ways in which human behaviour may accentuate the severity of 'natural' occurrences, for instance by intensive development of flood plains and liquefaction-prone land (Hewitt 1997; Pawson 2011). Furthermore, impacts are often spatially uneven and unpredictable. Vulnerability, therefore, is multidimensional with some aspects, notably the social, often more hidden from view than others (Cutter et al. 2003).

In the geographical literature, vulnerability is frequently juxtaposed with resilience. If vulnerability is the susceptibility to be harmed, then resilience refers to the magnitude of disturbance that can be absorbed before radical change occurs in a social or ecological system. It is used as well to describe the capacity to cope with emerging circumstances (Adger 2006). From origins in materials science and engineering, it has been picked up in ecology and social science, to include the study of how communities and social networks respond after disasters (Blair and Mabee 2020; Wilson 2012). It has also become a popular research concept in studies of climate change and development. Mainstream views of resilience however tend to privilege the persistence of a system over its transformation, whereas it is important to question what sort of resilience, and for whom (Hayward 2013; Tanner et al. 2015)? Its widespread adoption in policy since the neoliberal revolution of the 1980s, promoting the self-reliance of 'resilient' individuals and communities, appeals to those who promote a smaller, less engaged state (Joseph 2013). Furthermore, resilience approaches may damage 'practical understanding of the causative processes of vulnerability and how disasters come to be created' (Weichselgartner and Kelman 2015, 259).

Nonetheless, despite the pervasive practice of forms of neoliberalism in New Zealand (Le Heron and Pawson 1996), earlier traditions of collective security persist. These originated in the desire for a strong national vision based on full employment, provision of housing and social welfare (Easton 2000; Sutch 1966). This included old-age support, which has become politically entrenched, along with the later no-fault Accident Compensation Corporation (ACC) scheme, dating from 1972, which affords collective cover for all forms of personal injury. Most historical accounts, however, overlook the introduction of security against disaster, initially providing for compulsory insurance of property against war damage in 1941 but, after a destructive series of earthquakes in the lower North Island the following year, extended in 1944 to also include earthquake damage. Other types of disasters were progressively added in the post-war years (Department of Statistics 1990). The Earthquake and War Damage Commission became the Earthquake Commission in 1993, providing cover, in spite of its name, against a wide range of environmental hazards. The Commission provides capped insurance for residential buildings as a compulsory element of private house policies; damage over the cap is the responsibility of the private insurer. Homes without private insurance are not eligible, nor are commercial properties (DPMC 2017).

Like almost all settlements in New Zealand, Christchurch is vulnerable to a range of environmental hazards, including severe floods in 2014 and the Port Hills fires of 2017 (Uekusa et al. 2022). However, before 2010, public perception of the likelihood of future earthquakes was low (McClure et al. 2011). Schoolchildren in New Zealand practice earthquake drill, but the risk was not well appreciated *here*. Local earthquake tremors were not unknown, but harm was unusual and distant in time: a significant event located beneath New Brighton in 1869 in the early days of colonial settlement is lost to social memory (Elder et al. 1991). The Anglican cathedral spire was toppled not long after completion, in 1888 and again in 1901, by earthquakes in north Canterbury. On both occasions it was replaced, surviving until the spire and upper half of the tower was destroyed over a century later on 22 February 2011. Lack of awareness of the proximity of seismic danger was therefore a significant factor in engendering a sense of security in the city, even in an earthquake-prone land. But this had also long been reinforced by powerful technocratic instruments, in the form of substantial insurance cover and the requirements of a building code. These are considered next.

Technocratic security: insurance and building codes

Globally, the risk of earthquakes in urban areas is especially great around the Pacific rim, with the proto-typical examples of public sector involvement in earthquake insurance markets being in the high-income locations of Japan, California and New Zealand. Given the particularities of New Zealand's quest for security, the Canterbury earthquake sequence 'turned out to have been the most heavily insured earthquake event in history' (Nguyen and Noy 2017, 1), as well as the second biggest global insurance loss from that cause. The largest to date has been the Honshu earthquake and tsunami of March 2011, but the extent of insurance loss relative to the value of residential and commercial assets was most significant in Canterbury. Table 2.1 reveals that the proportion of economic losses covered by insurance for the three biggest episodes in the Canterbury earthquake sequence (on 22 February 2011, 4 September 2010 and 13 June 2011) was significantly higher than in large Japanese or Californian events, or those elsewhere listed amongst the world's ten costliest insured earthquakes. The table is drawn from a report by the international insurer, Marsh, three of whose staff were killed in its central

Table 2.1 World's costliest insured earthquakes

Date	Location	Deaths/ missing	Insured losses	Economic losses
11 March 2011	Japan	19 135	US$35.7 billion	US$210 billion
22 Feb 2011	**Canterbury, NZ**	**185**	**US$15.3 billion**	**US$23 billion**
17 Jan 1994	California, US	61	US$15.3 billion	US$44 billion
27 Feb 2010	Chile	562	US$8.4 billion	US$30 billion
4 Sept 2010	**Canterbury, NZ**	-	**US$5 billion**	**US$6.5 billion**
17 Jan 1995	Japan	6 430	US$3 billion	US$100 billion
29 May 2012	Italy	18	US$1.6 billion	US$16 billion
26 Dec 2004	Indonesia	c. 220 000	US$1 billion	US$11.2 billion
17 Oct 1989	California, US	68	US$960 million	US$10 billion
13 June 2011	**Canterbury, NZ**	-	**US$800 million**	**US$2 billion**

Christchurch office on the top of the Pyne Gould Corporation building when it collapsed on 22 February 2011 (Marsh 2014).

The reason is that Earthquake Commission cover is significantly more affordable than other international earthquake programmes, and is levied at a flat rate regardless of evaluated risk. It differs in this respect from those of the Japanese government and the California Earthquake Authority. The Japanese scheme provides a sum insured of only 30–50 per cent, and take-up is below 30 per cent. California's earthquake insurance is expensive coupled with relatively high levels of excess: only about 10 per cent of households participate. In Christchurch, almost all residential properties had private insurance cover, so were eligible for Earthquake Commission help. Its aggregate spending in relation to the Canterbury earthquake sequence reached about $NZ11 billion by 2016, with private insurers paying an additional $NZ4.3 billion. Modelling suggests that for 'a similar size disaster event' in Japan and California, their insurers would have spent only $NZ2.5 and $NZ1.4 billion, respectively (Nguyen and Noy 2017, 16). Noy et al. (2017, 18) describe the New Zealand scheme as 'the one successful outlier' internationally.

This comment implies that the Earthquake Commission model provided the security that citizens sought, but in reality, the outcome was quite different. The Earthquake Commission Act 1993 had 'not envisaged a sequence of events that would result in successive (and compounding) losses under single household policies' (DPMC 2017, 11). The Commission was quickly overwhelmed by the volume of claims, amounting to over 700,000 by March 2012 in respect of 160,000 damaged residential properties (about 90 per cent of the total) in and around Christchurch. This was five times larger than the Commission's prior expectation of the worst foreseeable event (King et al. 2014; also DPMC 2017; Potter et al. 2015). Government's subsequent requirement that all insurers, including the Commission, offer a repair and rebuild settlement, rather than cash payments, did nothing to speed up the process. Marsh (2014, 13), in comparing outcomes from the February 2010 earthquake in Chile, the February 2011 occurrence in Christchurch and that in March 2011 in Japan, comments that 'New Zealand was the least prepared of [the three] from an insurance perspective' (cf. Grollimund 2014). In a practical sense, this could be attributed to the fact that it had been 70 years since an earthquake had significantly affected a major urban centre.

The Earthquake Commission therefore had to quickly increase its staff from 22 to a peak of about 2200 to handle the volume of claims (King et al. 2014). There was a chronic shortage of experienced loss adjusters, as well as of building and engineering expertise. Home owner challenges to the size of settlements offered by both the Commission and private insurers proliferated, delays to settlements often took years, and even then shoddy repairs were legion (Greene 2021). A Public Inquiry was established in 2018 to establish why a landmark social insurance scheme, designed – or so it was widely believed – to secure and protect citizens from material and emotional vulnerability in the presence of earthquakes, had done anything but. The inquiry concluded that staff of the Commission 'on the whole' did their best in difficult circumstances, 'even if it may not have appeared that way to the public at the time' (Public Inquiry 2020, 7). This is the other side of the story illustrated in Table 2.1. It contrasts with Japan, where the experience of the Kobe earthquake in 1995 helped with disaster planning, such that 99 per cent of claims from the Honshu disaster were settled within 11 months (Marsh 2014; Noy et al. 2017).

For New Zealand, however, another important facet of the tension between vulnerability and security lies in how different stakeholders understand the country's building code. The code has its roots after the Hawkes Bay earthquake destroyed much of the towns of Napier and Hastings in 1931. Recommendations were made about standards of design and construction, so that buildings would better resist horizontal ground motions. Masonry structures had to be firmly bonded to move as one unit. The code was frequently revised to accommodate subsequent changes in building materials and design. There were major amendments between the late 1960s and early 1980s, with the introduction of ductility standards (how well a building will endure lateral displacements imposed by ground shaking) and overall capacity design. The code is a 'high-level performance-based document that defines the overall objectives, functional requirements and performance requirements for buildings' (Department of Building and Housing 2012, 7). It does not require specific materials, design or construction methods. The aim in a moderate earthquake is to protect the building from damage; in a severe earthquake, it is to protect life.

In the wake of the Canterbury earthquakes, structural engineers felt that there was little public understanding of this point, there being 'a disconnect between what engineers understand as "earthquake-resistant" and what clients and tenants perceive as "earthquake-resistant"' (Hare 2011, 2). This was one reason for the establishment of the Canterbury Earthquakes Royal Commission in 2011, with a focus on why two modern buildings in the city centre collapsed with considerable loss of life, and why a number of others of recent vintage were badly damaged. It is evident that there were considerable differences in expectations, as well as in interpretations of earthquakes as events. Many members of the public came to rely on estimates of earthquake severity in terms of magnitude on the Richter scale, as communicated by the Earthquake Commission's popular Geo-Net website. Structural engineers however use shaking intensity (measured on the Modified Mercalli scale) or ground acceleration. Furthermore, code requirements are minimum standards, not target levels, and are obviously social constructs. Hence the saying that 'earthquakes do not read the code': the shaking intensity on 22 February was such that 'the seismic loading was at least twice the design level for normal use buildings' (Kam et al. 2011, 272).

An important plank in the state's response to the findings of the Royal Commission (2012a) has been the Building (Earthquake-prone Buildings) Amendment Act 2016. This categorises New Zealand into three broad seismic risk areas (high, medium and low), and introduces specific timeframes for identifying and remediating buildings within each. It therefore aims to provide a more effective and consistent framework for identifying earthquake-prone buildings around the country, to better target areas of greatest risk, as well as a methodology for assessing the seismic capacity of buildings (Ministry of Business, Innovation and Employment 2017). That Canterbury would be placed in the highest risk category might have surprised many of its inhabitants before September 2010 (cf. McClure et al. 2011).

The country's first state of the environment report for example stated that 'Compared to other parts of the earthquake-prone Pacific rim … New Zealand has a moderate frequency of earthquakes, with most no greater than 4 on the Richter scale' (Ministry for the Environment 1997, 8–15). This was illustrated by a map relating earthquakes to the line of the major Alpine Fault, which lies on the other side of the Southern Alps to Canterbury and Christchurch. New Zealand *Official Yearbooks* have been more circumspect, describing a pattern of shallow quakes which are 'responsible for almost all damage to property, and are widely scattered throughout the country', before concluding that 'it would be highly imprudent to treat any part of New Zealand as completely free from the risk of serious earthquake damage' (Department of Statistics 1990, 6). In this spirit, engineering studies in the 1990s identified the risk to buildings and underground infrastructure in Canterbury from earthquake-induced liquefaction (CAE 1997; Elder et al. 1991). But this risk was not widely appreciated and liquefaction as a concept was more or less unknown to the public before the destruction to come.

The ecology of the earthquake sequence

Any complacency about the likelihood of serious seismic incidents in the region was punctured by the lengthy first tremor at 4.35 am on 4 September 2010; it was also noisy, compared by some to the sound of an express train (Rice 2011). But there was minimal social activity in the city that early in the day, saving widespread serious injuries and probable deaths. For example, if the collapse of the loaded bookshelves (all of which were freestanding) in the University of Canterbury's library tower had occurred during opening hours, there would have been a very different outcome. Radio reports from around the city described how the earth had come alive with sand blows, or 'mud volcanoes', ground fissures and lateral spreading cracks, especially around streams, and widespread surface liquefaction (Borella et al. 2020). Sodden masses of fine silt covered driveways and streets and even partially buried cars. Ground deformation occurred both vertically and horizontally (Hughes et al. 2015). Christchurch is by no means unusual and shares a susceptibility for liquefaction with many large cities in active seismic zones around the Pacific rim, and elsewhere, such as Jakarta, Tokyo, Vancouver, San Francisco and Los Angeles (Grollimund 2014). The extent of liquefaction however was uncommon, and unprecedented in New Zealand.

The earthquake-induced shear stresses that initiated this process originated from a shallow (~10 km) seismic event on the previously unrecognised Greendale fault about 40 kilometres west of the city. This fault is one of a network beneath the Canterbury Plains that are now known link to the boundary of the Indian and Pacific tectonic plates, marked through the South Island by the Alpine Fault. Prior to 2010, it was widely assumed that, in the unlikely event of an earthquake, the cause would be from movement of this fault (which is over 150 kilometres away), or of fault lines in north Canterbury (100 kilometres distant). The identified surface rupture of the Greendale Fault extends east for almost 30 kilometres towards the satellite town of Rolleston. This was considered a short distance for generating a Mw 7.1 earthquake, so it was initially hypothesised that much of the rupture was beneath the surface, but with no evidence of it continuing beneath the city (Cubrinovski et al. 2010).

Nonetheless, a continuing pattern of aftershocks rapidly emerged closer and closer to the city (Figure 2.1). There were 400 in the next week, with ten in excess of magnitude 5, one of which was centred near the port of Lyttelton. Another one of 5.0 in early October was near the western suburb of Templeton, producing aftershocks in the eastern part of the city under New Brighton and the Avon-Heathcote estuary. The duty seismologist at GNS Science, a Crown Research Institute, sought to reassure people that although big aftershocks were possible, albeit with decreasing probability over time, 'the general trend [is] for smaller and less frequent shakes. It's quietened down considerably over the last month' (in Gorman 2010a). But on Boxing Day 2010, there was a swarm of shallow shakes directly beneath the city, 'such that the ground has barely stopped shaking for more than a few minutes at a time' (Gorman 2010b). The biggest was magnitude 4.9, with many people feeling this more strongly than the initial 4 September event, given its proximity (Rice 2011).

This was the prelude to the dangerous earthquake of 22 February 2011. That event was deadly as it happened in the middle of a working day at 12.51 pm. The epicentre was near the port of Lyttelton, a few kilometres south of the city centre, and at a depth of only five kilometres. Thousands of people, among them many tourists, were displaced amongst damaged and collapsing buildings in the business district. In the west, the University of Canterbury's

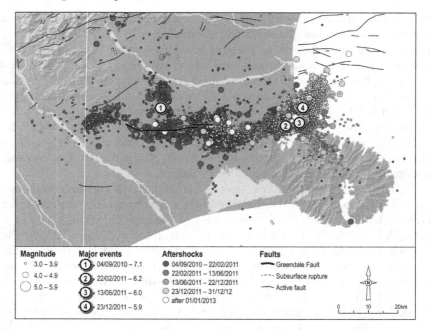

Figure 2.1 Location of epicentres in the Canterbury earthquake sequence, 2010–14.

Source: Earthquake data from Geonet (geonet.org.nz), capturing seismicity from 1 September 2010 to 31 December 2014. Fault lines from GNS Science Active Fault database (data.gns.cri.nz/af/)

campus was full of milling students on the second day of the academic year; the library, now with appropriately secured bookshelves, had just re-opened. In Brooklands, a coastal suburb, a driver described how 'the car started bouncing around heaps and throwing me around. I just … sat there watching all the liquefaction come up out of the ground. It was crazy and very scary' (*The Star* 2011, 12). Here, the water table is high, as in Kaiapoi which had been badly hit by liquefaction in September. Other places affected included eastern suburbs in the Ōtākaro Avon river corridor, and streets in the city near stream courses. Generally, the western parts of the city, on more stable gravels with a lower water table, were less affected. The spatial distribution of liquefaction (Figure 2.2) was quite similar to that anticipated in one of the earlier engineering studies (Elder et al. 1991).

If the overall extent of liquefaction was unprecedented in New Zealand, so too were other aspects of the February 2011 event. It did not fit in with the seismological prediction 'for smaller and less frequent shakes' over time. The eastern movement of seismicity had in effect displaced the release of pressure in the earth's crust amongst further previously unknown faults, triggering ongoing instability. But at magnitude 6.2, the February earthquake released far less energy than had been the case on 4 September. It was so damaging because the epicentre was shallow and under the city, even though the initial tremor lasted only ten seconds or so. Peak ground force acceleration recorded

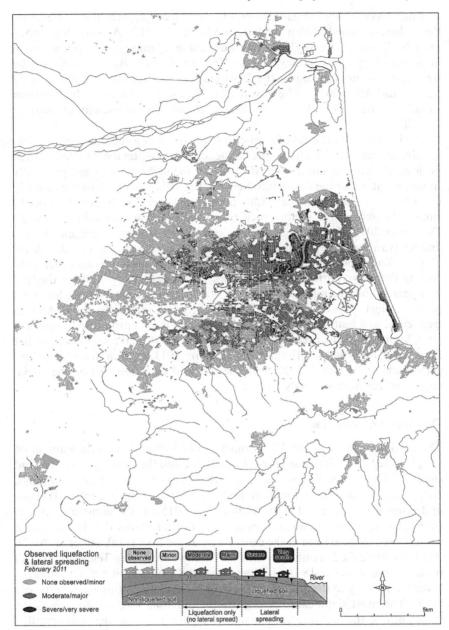

Figure 2.2 Extent of liquefaction on 22 February 2011.

Source: LINZ (2015), redrawn from Appendix A, map 4b.

that day was initially estimated at more than 2G, or twice the force of gravity, the highest recorded in New Zealand (Kaiser et al. 2012). A seismologist with the US Geological Survey claimed would make it 'among the most powerful in terms of ground-shaking acceleration on record'. By comparison, the strongest ground shaking in the 2010 Haitian earthquake was about 0.5G (Lin II and Allen 2011). There were also extensive rockfalls in the southern suburbs in the Port Hills, and dramatic collapse of both present and former sea cliffs.

The February earthquake is sometimes regarded as a separate episode, but usually as part of the aftershock sequence from September. Either way, it prolonged that sequence, with aftershocks throughout the year, gradually trending out under the ocean into Pegasus Bay (Figure 2.1). Two particularly large tremors under the Port Hills, both at magnitude 5.6, occurred on 13 June 2011, setting off further rock falls in hill and seaside suburbs. On 23 December 2011, there were two more substantial shakes of magnitudes 5.8 and 5.9 (van Ballegooy et al. 2014). The overall pattern after September 2010, along different previously unrevealed faults, was of seismic events each gen-erating their own pattern of aftershocks. There were over 12,000 in the first two years, 3,600 being of magnitude 3 or greater, strong enough for most people to feel. This continued in subsequent years, gradually diminishing in frequency, as Figure 2.1 shows. For this reason, it is appropriate to refer (as per Chapter 1) to the event of 'the Canterbury earthquake sequence', rather than 'the' Canterbury earthquake of February 2011, or even 'the Canterbury earthquakes of 2010–11'. Neither the physical occurrences nor the social consequences were so neatly contained.

Death and destruction

The headline feature of the 22 February 2011 earthquake was the number of people killed, given its occurrence at lunchtime and the severity of the ground shaking. There was also widespread destruction of central city buildings, serious impairment of a range of underground and surface infrastructure, and extensive damage to homes (Potter et al. 2015), as the number of claims to the Earthquake Commission reveals. The death toll was 185, 169 of whom were known to have been in the central city at the time (McLean et al. 2012). But 133 people died at only two sites: 115 in the Canterbury Television (CTV) building and another 18 in the Pyne Gould Corporation (PGC) building. Claims for injury to the Accident Compensation Corporation came to nearly 7,000. The most significant reason why the outcome was not worse was the stringency of the building code designed, as discussed above, to protect life. By comparison, the mortality rate for the Kobe earthquake in Japan in 1995 was 30 times higher, despite it happening in the early hours of the morning and with lower peak ground accelerations of 0.8G (Ardagh and Deely 2018). Standing behind the Canterbury statistics however is the emotional turmoil and disempowerment stemming from coping with endless, unpredictable shocks, quite apart from ongoing contests with insurers (Hayward 2013;

Spittlehouse et al. 2014). The Prime Minister's Chief Science Advisor drew attention to the long-term psychosocial impacts of the earthquake sequence as early as May (Gluckman 2011).

Putting the impacts on human life and health to one side for the moment, the damage wrought to structures was considerable. But 'structures' are also significant experiential elements of people's lives, as homes, workplaces, social facilities and places of worship. One illustration of how widely the effects of the September 2010 earthquake were felt in homes across the city is a striking map produced from Earthquake Commission data. This locates the more than 30,000 claims for chimney or fireplace damage received by late January 2011, that is before the devastating event on 22 February that year (Figure 2.3). The pattern is fairly evenly spread across the urban area except in the newer, outer suburbs where fewer houses had chimneys. Elsewhere, brick chimneys were usually the most unyielding part of the city's typically light timber-framed homes. Many collapsed in the September earthquake,

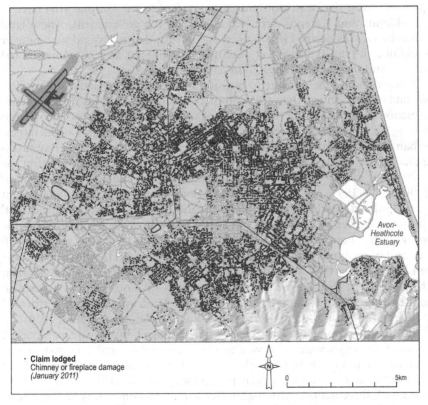

Figure 2.3 Location of claims for chimney or fireplace damage lodged with the Earthquake Commission by late January 2011.

Source: Redrawn from Rice (2011, 91).

shearing off at the roofline, sometimes falling through the roof. Others cracked nearer the ground and also needed to be removed (Rice 2011).

Liquefaction-induced damage to houses depended on location and construction type. Older homes built on timber piles, as well as more modern ones on concrete slabs, often tilted, sank or settled differentially. Lateral spreading compounded these effects (Cubrinovski et al. 2010). Sometimes the concrete slabs fractured, steel reinforcement having not previously been a requirement. Older piled houses with recent extensions on slabs suffered extensive damage at the interface (Buchanan and Newcombe 2010). Shaking-induced damage was variable, with masonry veneers more vulnerable than wooden weatherboards. Cracking to internal wall and ceiling linings was widespread, although '[l]ight timber framed buildings provided excellent life safety' (Buchanan et al. 2011, 357). There were considerable differences at street level, reflecting soil conditions as well as house type and construction. Hill suburbs were prone to rockfall and severe ground shaking in the major 2011 quakes; homes with heavy tile roofs, rather than the more common iron or colour steel sheeting, were often worst affected. Parts of the hills, the Ōtākaro Avon river corridor and land adjacent to the Waimakariri river in Brooklands and Kaiapoi were 'red zoned' by the government, where land remediation was deemed impractical due to subsidence, ground weakness or rockfall risk (van Ballegooy et al. 2014). The process and outcomes of red zone abandonment are discussed in later chapters.

Compounding the damage to housing was the state of streets and underground infrastructure. Over 1000 kilometres of urban roadway needed reconstruction (over half of the city's total). Many less damaged streets also had to be dug up to access compromised underground services. This was the largest urban lifelines disruption in a New Zealand city since the Hawkes Bay earthquake of 1931, but the impact was reduced due to work done following the assessment of the vulnerability of those lifelines in the city since the mid-1990s (CAE 1997). In 2002, the Civil Defence and Emergency Management Act stipulated that lifelines be able to function to 'the fullest possible extent' during and after an emergency. Seismic protection work undertaken by the local electric power distribution company enabled it to restore service to 90 per cent of customers within ten days of 22 February 2011, although at a cost ten times greater than after the September 2010 and June 2011 quakes, when the 90 per cent level was achieved within a day. Previous lifelines' work also ensured that the 180-kilometre liquefied petroleum gas network performed well, and far better than in overseas earthquakes (Giovinazzi et al. 2014).

Both power networks are owned and managed by private companies, whereas the city's water and wastewater networks are publicly owned and managed. Parts of these networks are considerably older than those of the utilities, made of earthenware clay pipes susceptible to breakages, which then filled with liquefaction. Although water supply to homes was restored relatively quickly after the September 2010 earthquake, this took more than a month after the February and June 2011 episodes for properties in severely compromised districts. Wastewater pipes were similarly badly affected and by

August 2011, about 800 houses still required repairs to their private sewer pipes to be fully functional (Giovinazzi et al. 2014). In the weeks after 22 February 2011, the city council distributed over 40,000 chemical toilets to households in need, mostly in the eastern suburbs. More than 2,000 portable toilets ('portaloos') were placed on streets, although there was much frustration over the equity of distribution (Heather 2011a; Potangaroa et al. 2011). Months of intrusive disruption to daily routines for those people in the worst affected areas only added to emotional stress.

Although very real, damage to homes and suburbs was less dramatic than that to the city centre. Here seismic shaking 'significantly exceeded the 500-year return period design level' of the current building code (Kam et al. 2011, 240). The age and construction of buildings were key factors in their fate. Almost half of the total central city building stock was built prior to the introduction of modern seismic codes in the 1970s, although this included a lot of residential buildings of mixed ages, as the inner area of the city is very spread out. But the risk posed by older commercial and public central city buildings had long been recognised. In a 1996 TV programme called 'Earthquakes!', the city council's then buildings engineer said to the camera 'Your heritage list, to be quite cruel, defines most of your earthquake-prone buildings'. He pointed to problems with older two or three-storey structures in the main retail area. A lot of dangerous parapets had been removed, but the standards to which such buildings were held were a fraction of those applying to more modern ones (Dixon Productions 2011).

Some significant masonry heritage buildings had however been seismically strengthened, although with differing outcomes. For example, the Victorian Gothic Canterbury Museum survived in February 2011 reasonably unscathed, but not the Anglican Cathedral, built in the same style and period. Others that had not been strengthened at all, such as the remarkable stone chamber of the Provincial Government Buildings of 1865, collapsed completely, as did the west end of the neoclassical Catholic Cathedral (finished in 1906), which had been partially strengthened. These were perhaps the two finest heritage buildings in the city (Ansley 2011). Old masonry facades on retail streets that were barely connected to the Victorian brick boxes behind them also disintegrated: eight people died when they were crushed in buses beneath falling masonry. The two deadliest buildings however were products of the 1960s and 1980s, respectively: the PGC and CTV buildings.

The Royal Commission (2012a) was charged with investigating why some buildings failed totally and why buildings differed in the extent of failure. Volumes 2 and 6 of its final report, which cover the PGC and CTV cases, are particularly valuable from a social perspective as they contain accounts from those who were rescued and tributes for those who died. The PGC building was designed to codes of practice used in the 1960s, before ductile performance was a consideration. On 22 February, the building underwent shaking 'that was almost certainly several times more intense than the capacity of the structure to resist it'. But 'neither foundation instability nor liquefaction was a factor in the collapse' (Royal Commission 2012b, 38). David Sandeman, an

Figure 2.4 David Sandeman and Jeff McLay celebrating their rescue from the PGC
building on 22 February 2011.

Credit: Stuff Limited.

employee of Marsh, was on the fourth floor when the tremor started and
found himself 'plunging down. I estimate it was about forty feet as we ended
up on the first floor as I subsequently discovered' (Royal Commission 2012b,
36). The photograph of his later rescue alongside a colleague became one of
the defining images of that day (Figure 2.4).

Like the PGC building, the CTV building had multiple tenancies, which
included (apart from Canterbury Television) health providers and an English
language school. This is why a large number of overseas nationals lost their
lives there. The Royal Commission pointed to 'a building design that was
deficient in a number of important respects' (2012c, 302), but again found
that ground conditions were not a factor. It concluded that a building permit should not have been issued 'because the design did not comply with the
[city council's] building bylaw' (2012c, 303). The building pancaked within
10 to 20 seconds of the onset of the earthquake, followed by a fire that
burned for some days. 'It was a sudden and catastrophic collapse' (2012c,
307), the most acute example of building failure during the extreme ground
motions of the 22 February earthquake. But many other buildings, even if
they did not cost lives, were too seriously compromised that day to remain,
or became uneconomic to repair once the extent of damage became apparent (Marsh 2014). Eventually, as shown in Chapter 6, about 70 per cent of
the central city's building stock had been demolished (Bennett et al. 2012;
McLean et al. 2012).

Rapid disaster response

The scale of the disaster and need for resources mobilised at and beyond the national level prompted a Declaration of National Emergency on 23 February 2011, the first time this measure had been used under the Civil Defence and Emergency Management Act of 2002 (DPMC 2017). It remained the only such declaration until 25 March 2020 at the onset of the coronavirus epidemic. Response to the September 2010 earthquake had been managed as a 'regional emergency'; most civil defence emergencies, which in New Zealand usually concern flooding or severe weather, are declared at local or regional levels (Civil Defence 2020). Although such declarations are civil defence management tools, many other organisations, groups and individuals were involved in rapid response in the hours, days and weeks that followed the major episodes in the Canterbury earthquake sequence.

The headquarters for the Civil Defence operation, or 'Christchurch Response Centre', was established in the city's relatively new downtown Art Gallery (completed in 2003). This served as the coordination hub for the response for up to 500 personnel until it was necessary to move two months later due to damage to the building. The emergency response effort initially focused on the city centre as it was the locus of death, injury and visible building damage. The subsequent civil defence review identified a lack of 'situational awareness' at the Response Centre about what was happening elsewhere around the city, a charge echoed from the Ngāi Tahu tribal organisation (Solomon 2021). It also recognised the need for 'clear and unambiguous' identification in emergency planning of lines of responsibility for incident control at particular disaster sites (McLean et al. 2012, 39). This was the consequence of the range of emergency services involved, including police, fire and ambulance, as well as volunteers, all soon joined by urban search and rescue teams (USAR) from around the country and overseas, as well as defence force personnel. It was fortuitous that at the time there was a major joint defence training exercise underway around the city, as part of which the naval vessel HMNZS Canterbury was in port at Lyttelton.

Christchurch hospital however did have a clear major incident plan, covering both internal incidents (those that impair the ability to provide the services expected) and external incidents (which generate overwhelming demand for those services). As the building was damaged, and there were immediate needs to overcome issues with power, water and sewerage provision, managers oversaw the internal incident response whilst emergency department medical staff attended to the external incident. The President of the Royal College of Surgeons of Edinburgh, who witnessed this (having been in the city for a conference, and attending the hospital with his injured wife), subsequently wrote that 'I am in little doubt that many more people would have died in the hospital that day without the skill and leadership displayed in the Emergency Department' (in Ardagh and Deely 2018, 82). After an initial rush of patients, arrivals slowed as those who had been released from being trapped in damaged buildings trickled in.

The search for survivors in damaged and collapsed buildings was carried out at first by emergency services and members of the public, until joined by the search and rescue teams. New Zealand had three such specialist teams, based in Christchurch, Palmerston North and Auckland, with about 50 personnel each from the fire service, and including medics, engineers and specialist search dogs. They were joined by international USAR teams from Australia, China, Japan, Taiwan, Singapore, the United States and the United Kingdom (McLean et al. 2012). Their origins reflected affective ties, countries in earthquake-prone areas with the necessary capabilities, and the places that people trapped in collapsed buildings were known to be from. The Japanese team, which like those from Australia and Singapore, also brought search dogs, was allocated to the CTV building as there were many Japanese students in the language school that had occupied level 4. In addition, 20 trained volunteer response teams from Christchurch and around the country joined the search and rescue operation. Some of these had also deployed during the September earthquake, when one of their tasks, apart from carrying out initial building inspections and cordoning unsafe areas, was to assist in the dismantling of residential chimneys (Seager and Donnell 2013; Figure 2.3).

Within a few hours of the 22 February event, a cordon was established around the central business district by members of the Defence Force and police, to protect people from falling debris and to secure the area once vacated. The cordon was initially set up with manned checkpoints at road access points, but within days a continuous physical barrier had been erected (illustrated in Chapter 6). Defence force members were to remain at the cordon for more than a year; the cordon itself lasted far longer, although shrinking over time as city blocks were rendered safe through demolition and clearance. In the port town of Lyttelton, which was very close to the epicentre of the earthquake and had many older masonry buildings, the response was led by local community groups, underpinned by the resources of HMNZS Canterbury, which provided meals for the 1000 people left homeless in the town (McLean et al. 2012). Coincidentally, during the 1931 Hawkes Bay earthquake, it was sailors from HMS Veronica – a ship in the New Zealand Division of the Royal Navy – who had provided much of the early response on the ground when in port at Napier (Wright 2001).

The subsequent civil defence response review briefly records the ways in which social organisations, the police and spontaneous volunteer groups worked together in the worst affected suburbs. 'Overall', it concluded, 'the resilience of the community was remarkable' (McLean et al. 2012, 175). The Lyttelton example was not alone. The south-eastern coastal suburbs of Redcliffs and Sumner, effectively isolated by rockfalls, self-organised with the volunteer fire brigade and volunteer coastguard taking the lead, supported by local police. In the review they stated that they 'had practically no contact with the Christchurch Response Centre until some weeks after the event' (McLean et al. 2012, 182); similar experiences were reported from other hill and harbour suburbs. In the eastern coastal suburb of New Brighton, the response was led by local police with 'some 60,000 people ... fed out of local community-led distribution centres' (McLean et al. 2012, 71), aided by members of the defence force, the Farmy Army and the Student Volunteer Army.

If one feature above all came to symbolise the Canterbury earthquake response, it was the thousands of volunteers who joined the Farmy and Student Volunteer Armies. Both initially formed at the time of the September 2010 earthquake: the first by Federated Farmers, as rural as well as urban areas were badly affected in that event; the second by university students (Hayward 2013). One of their major roles then and in February was in clearing liquefaction, particularly around people's homes, checking on residents' welfare and distributing food. The Student Army was also tasked with delivering information sheets in affected suburbs, and assembling portaloos for placement in streets (which arrived as flat packs from the United States in early March). The review observed that civil defence planning had not taken into account how to embed such large-scale volunteering, but that 'a key learning was that ... management must be left with the [volunteer] group' (McLean et al. 2012, 182). It also observed that next to no use was made of social media in the emergency response, with the exception of the Student Army, which was mobilised every time there was a need through Facebook. The social significance of this organisation in earthquake recovery is assessed in Chapter 7.

Conclusion

This chapter has outlined some of the key contours of the Canterbury earthquake sequence as a socio-seismic event series. In strictly seismic terms, none of the tremors were especially large, but globally they are amongst the costliest earthquake events (Table 2.1), due to their proximity to the city of Christchurch, and very high levels of residential and commercial insurance. The sequence also did not accord with scientific expectations of a major shock, followed by a diminishing, if erratic series of aftershocks. Instead, it was prolonged as pressure release in the earth's crust was displaced eastward beneath the city through a previously unrecognised network of faults. A great deal of damage was caused in the city centre by high ground-shaking acceleration during the quake of 22 February 2011, even though this was quite short in duration. Elsewhere in the city, but especially in the eastern suburbs, in each major event, the ground came alive as liquefied materials were ejected from beneath the surface, smashing foundations and underground infrastructure.

A number of significant social themes have been highlighted. Ironically, this turned out to be a city that has been far more vulnerable than the technocratic framework of property insurance and building codes led many to expect. In fact, in terms of the model of recovery presented in Figure 1.2, insurance disputes have often greatly extended the trauma of the original quakes. And even if the potential for liquefaction was not widely appreciated, it was known about; those utility providers that had acted on this knowledge were able to restore power and communications networks quickly. Likewise, elements of the first response were well prepared, even if a lack of situational awareness at the Christchurch Response Centre meant that community networks, civic organisations and spontaneous volunteer mass mobilisation had to take ownership of recovery in different places. The extent to which 'the resilience of the community was remarkable', as the later Civil Defence review claimed, is a theme addressed in the next chapter.

3 Impacts on households and communities

Introduction

The earthquakes played out across a socially uneven urban landscape, affecting neighbourhoods with differing levels of material, financial and political resources. Within this context, this chapter considers the impacts on households and communities. It outlines some common aspects of the immediate experience of the quakes, then focuses on key processes that ensued, such as departures (where residents voluntarily or involuntarily left their homes and neighbourhoods, for shorter or longer periods) as well as the compensating arrivals of migrant workers to assist with the city's repair and reconstruction. The land zoning decisions and insurance settlements are then examined, with their impacts illustrated through a consideration of residents' experiences in the coastal suburb of Southshore. This leads to a consideration of psychosocial well-being, and the increased incidence of anxiety and depression across the city, which occurred to differing degrees for different groups. The quakes also prompted novel expressions of mutual aid at the community and neighbourhood level as people reached out to assist others. Along with broader welfare initiatives across the city, these expressions of support and care generally promoted recovery, as well as contributing to local resilience.

A differentiated urban landscape: patterns and movements

As is common to most modern cities, the urban landscape of Christchurch is socially and economically differentiated. In broad terms, there is a northwest-to-southeast gradient of affluence, with neighbourhoods in the west and northwest of the city having higher than average household incomes and property values. House prices and household income generally decrease as one moves south and east from the city centre. This trend continues until the hill suburbs (eg. Cashmere and Mt Pleasant) are reached, where household incomes and housing values increase, gradually around the bottom of the hills and then more rapidly on the hills themselves. The resulting socio-economic differentiation of the urban landscape can be seen in the patterning of social deprivation. Figure 3.1, which represents census area units in terms of their average score for the 2018 New Zealand Deprivation index, a composite

DOI: 10.4324/9780429275562-4

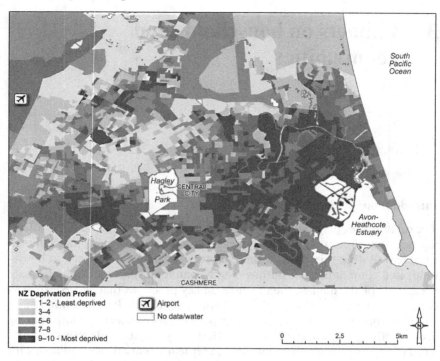

Figure 3.1 Socio-economic deprivation in Christchurch.

Source: https://ehinz.ac.nz/indicators/population-vulnerability/socioeconomic-deprivation-profile/

indicator of social deprivation, is indicative. Lighter tones here indicate more affluent areas while the darker tones depict more deprived areas. The concentration of relative deprivation in the eastern suburbs is evident.

The earthquakes intersected with this patterned landscape in ways that at times compounded already existing socio-economic disadvantage. Although residential damage occurred across the city, homes and buildings in the west and northwest – except those close to streams – were generally less affected than those in the east, where more extensive liquefaction and lateral spreading damaged both houses and subsurface infrastructure. Many eastern neighbourhoods had above-average levels of social deprivation, and the disruption of the earthquakes often contributed to additional hardship. As has been evident in other post-disaster cities (eg. New Orleans), a number of these eastern communities subsequently felt neglected by government recovery organisations, to the extent that they formed a range of protest and advocacy groups (as discussed in Chapter 10). In the hill suburbs, a significant number of properties suffered damage from shaking and some were impacted by rockfall. In financial terms, however, many of these households were relatively affluent, with their owners well positioned financially to find alternative accommodation during the repair or reconstruction of their homes. Their financial resources assisted them to deal with the impacts of material loss.

As signalled in the previous chapter, the 22 February 2011 earthquake was experienced in diverse ways, and of all the episodes in the Canterbury earthquake sequence, it was the most significant given its daytime occurrence and location beneath the city itself. In the city centre, the initial event and aftershocks were strongly felt, and many people sought to shelter – if they could – underneath suitable protective structures. After the intensity of shaking subsided, workers spilled out onto the streets, keeping their distance from potentially precarious masonry facades, and taking in what had happened. When two of the authors walked into the central city that afternoon, they were met by hundreds of people leaving on foot, including many in business attire, along with tourists and school students. Some individuals appeared shocked. Most were quiet and seemed focused, presumably on getting home. A few were talking to each other. In the centre itself, there was a considerable amount of dust from damaged buildings and cracked roadways. Intermittent car and building alarms could be heard, mixing with the distant sirens of emergency services vehicles. Alongside the tragic deaths and thousands of physical injuries, a number of people were successfully rescued that day from damaged buildings. However, many would live with a degree of ongoing trauma as a result of their experiences (Hunt 2021a).

The usual rhythms of activity and connection in the city were abruptly altered that afternoon. With most traffic lights no longer functioning, many streets became progressively gridlocked as large numbers of people attempted to return to their homes by car. As the cellular telephone network had been damaged, voice calls generally were not possible and even text messages were not a reliable means of communication. Messages often arrived out of sequence, significantly delayed or not at all. This impaired people's ability to contact family and loved ones, heightening the sense of emergency and logistical disruption. Many people were only reunited with their families and neighbours several hours later that day.

There were also longer-term population movements, however, particularly in response to the 22 February quake. In the weeks following, large numbers of individuals and households left Christchurch, either temporarily (for relief from aftershocks, or while short-term repairs were made on their homes) or on a longer-term basis (if substantive repairs or rebuilding were required, or if the circumstances were simply too overwhelming). The water and sewage systems in parts of the city were heavily damaged, and many residents in the eastern suburbs in particular lacked access to potable water or a working toilet for weeks if not months. Around 15 per cent (55,000 people) of the city's population is estimated to have left in the first week after 22 February, with women and families with young children over-represented in these departures (Newell et al. 2012). Postal redirection data also indicates substantial medium-term population movements in the immediate aftermath, with around 20,000 households redirecting their mail to other addresses within the city, and around 5,000 households to an address outside of Christchurch (Price 2011). The highest levels of long-term outmigration were from the eastern suburbs (Begg et al. 2021) – including neighbourhoods such as Avonside, Dallington and Richmond – where many houses had been rendered uninhabitable or were later subject to 'red zoning' (a government process of encouraged departure, as elaborated below). The majority of those departing moved elsewhere in New Zealand or to places in Australia (Adams-Hutcheson 2015; Campbell 2014; He et al. 2021).

In the 2013 census, the population of the Christchurch city was 341,469. This was an approximately two per cent decrease from the 2006 census population of 348,456, but a more significant 8.4 per cent fall (31,131) from the June 2009 estimate of 372,600 (Christchurch City Council 2010a). By 2018, however, the city's population had risen to 369,006, representing an increase of 8.1 per cent on the 2013 figure. A sizeable proportion of this growth reflected an influx of migrant workers that arrived in response to employment opportunities in the construction and building industries (How and Kerr 2019). Irish, English, Filipino and Australian workers were favoured by rebuild contractors, and in 2014, the majority of migrant workers in Christchurch came from the United Kingdom (25 per cent), Ireland (25 per cent) and the Philippines (20 per cent) (How and Kerr 2019; Searle et al. 2015). Given the extensive damage to the city centre and urban environment more generally, however, domestic and international tourism to the city fell significantly, and there were significant job losses among accommodation providers, cafes and restaurants (How and Kerr 2019).

Land zoning and insurance claims

As people grappled with the immediate material and psychosocial impacts of the February 2011 earthquake, central government was developing its recovery strategy. Elements of this strategy drew on the work of the Ministry of Civil Defence and Emergency in response to the September 2010 earthquake. In March 2011, central government established the Canterbury Earthquake Recovery Authority (CERA), whose responsibilities and remit were defined in the Canterbury Earthquake Recovery Act 2011. In essence, this organisation was tasked with overseeing and facilitating the recovery process on behalf of the Crown. With headquarters in Christchurch, one of its initial actions was to instruct a number of geotechnical engineering firms to undertake assessments of residential land across the city, so as to determine the extent of the damage from the February 2011 quake. In a sequence of announcements throughout 2011 and 2012, with the first on 24 June 2011, parcels of residential land across the city were designated as being either 'green zone' (suitable for ongoing occupation, and repair and rebuild of homes) or 'red zone' (unsuitable for ongoing occupation). In line with the criteria developed by CERA (New Zealand Government 2011), residential land was red-zoned if one or more of the following conditions held: there was significant and extensive area-wide land damage from liquefaction or rockfall; there was high risk of further damage to land and buildings from low-levels of shaking; the success of engineering solutions was considered to be uncertain and uneconomic; and any repair to the land would be disruptive and protracted.

Alongside these green and red zones – which encompassed the great majority of properties – the initial land zoning announcement in 2011 also included two temporary 'in process' categories. These were orange-zoned residential land, where further geotechnical and engineering assessment was needed, and white-zoned land, for hill properties where the risk of rock fall required further investigation. It took several years to complete the necessary further investigations, but by November 2014 the orange- and white-zoned land had all been re-categorised as either green or red zone.

The Crown expected houses on red-zoned land to be vacated, within a set of published timeframes, and facilitated by an offer of financial compensation. In contrast, households on green-zoned land were able to stay. The green-zoned land was divided into three 'technical categories': TC1 (where the land condition meant that house repair and rebuilding was considered unproblematic), TC2 (where repair and rebuilding could occur, although the land might have some liquefaction-related damage and settling) and TC3 (where repair and rebuilding was permitted, but site-specific geotechnical investigations were required, to determine what sort of foundations would be needed to compensate for the damaged land) (Christchurch City Council nd.). The TC3 land presented a challenge for homeowners whose properties needed extensive rebuilding or repair, as such work typically required either land remediation or, more commonly, deeper foundations, as a structural compensation for the compromised state of the land. The need for specialised and site-specific foundations typically added significantly to the cost of rebuilding, and there was some indication that this depressed the resale values of homes on TC3 land, as a form of zoning-related 'stigma' (McDonald 2013a).

Red-zoned land was concentrated along the Ōtākaro Avon river corridor, encompassing several eastern suburbs, as well as sections of several hill suburbs (both on the city and Lyttelton Harbour sides), and the western edge of the coastal suburb of Southshore (Figure 3.2). Significant areas of Kaiapoi, a small satellite town some 20 kilometres to the north of Christchurch, as well as coastal Brooklands, were also red-zoned. In view of the unsuitability of this red-zoned land for ongoing residential occupation – and given the Crown's liability through the Earthquake Commission for residential damage that might be caused by future earthquakes – the government made insured householders in these areas an offer of financial compensation in June 2011. The Crown would acquire either the house and land, as a package, or just the land (leaving owners to deal with selling or removing their houses independently). These offers were based on 2007/8 property valuations and owners were given nine months to decide whether or not they would accept the offer (CERA 2011; New Zealand Government 2011).

Responses to the Crown's red zone offers varied. Some homeowners contested the valuations, particularly when substantial home renovations or additions had been undertaken since 2007 (as the value of these improvements was not recognised in the financial offer). A proportion of red-zoned residents protested the more general requirement to surrender their homes

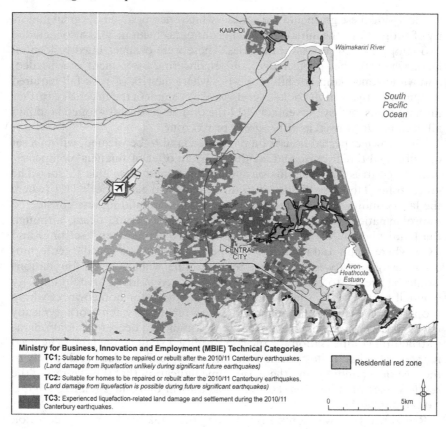

Figure 3.2 Land zones in Christchurch.

Sources: CERA, Environment Canterbury, Land Information New Zealand.

involuntarily, objecting to a central government process which they perceived as overriding their private property rights. Some homeowners also considered the risk assessment for their properties to be inflated or unreasonable (eg. in relation to vulnerability to rockfall in some of the hills suburbs), which amounted to disputing the validity of the technical assessment underlying the red zoning decision. Finally, a small number of households chose to stay on their land, declining to leave (Heather 2011b; Matthews 2012).

Despite these misgivings, by December 2015, 96 per cent (7,720) of the 8,060 red-zoned properties in Christchurch and nearby Kaiapoi had accepted the Crown offer (Nielsen 2016). These households vacated their properties and land by February 2016. From 2012 onwards, the Crown demolished houses and any outbuildings as they were vacated. A small number of houses were sold separately from the land, and then transported off-site (often after being cut into pieces to make this possible). This acquisition-demolition process generated extensive areas of open land in the suburbs along the Ōtākaro-Avon river, as well as smaller pockets in Kaiapoi, Brooklands and the hill suburbs. The appropriate ongoing uses of this red-zoned land then became the subject of sustained discussion and debate within the city (as discussed in Chapter 12).

Alongside the land zoning, the negotiation of insurance settlements was a central dimension of the experience of the earthquakes for many people in Christchurch. As noted in the previous chapter, New Zealand operates a dual public-private system of insurance against environmental disasters. All home-owners with a private insurance policy are automatically charged a small levy (a percentage of the annual policy premium), which contributes to the insur-ance funds for the government Earthquake Commission. When a house is damaged as a result of an environmental disaster – such as an earthquake, flood or fire – the government covers the first $NZ100,000 of the repair and rebuilding costs. Any work over this 'cap' becomes the responsibility of the private insurer. This dual scheme effectively means that the state takes on a portion of the risk associated with living in a seismically active country. For private sector insurers, the government's commitment also makes entry into New Zealand's relatively small market a less financially risky prospect. Although it has some benefits, the dual system also meant that many Christchurch earth-quake claims required complex multi-party negotiations between homeowners, the Earthquake Commission and private insurers (Miles 2016).

The timely and fair settlement of insurance claims became a central point of contention for many Christchurch residents (King et al. 2014; Miles 2016; Potter et al. 2015). Earthquake Commission and private insurer assessors often reached different conclusions about the extent of damage at a given property and thus the repairs needed. There is some evidence that such dis-putes were more common for claims around the $NZ100,000 boundary, when private insurers had the motivation to assess a claim as being below the 'cap', as the financial and also logistical responsibility for settling the claim would then fall to the Earthquake Commission (King et al. 2014). Many homeowners found themselves caught up in estimation disputes, after receiv-ing the Earthquake Commission or private insurance settlement offers that in their view under-estimated the damage and likely cost of repairs. Multiple rounds of assessment were sometimes undertaken at a single property, some-times by different assessors, who on occasion evaluated the scope of damage differently than their colleagues, introducing additional variability to an already complex process (Miles 2016).

Although many homes were eventually repaired using insurance proceeds from either government and/or private sources, there were complaints about the quality of the work undertaken, particularly when it was managed by the Earthquake Commission (with some repairs then having to be 'redone') (King et al. 2014; Poontirakul et al. 2017). Some homeowners believed con-struction companies were cutting corners during the building process in order to finish a project quickly and more profitably (He et al. 2021). The general shortage of suitable workers meant construction companies were at times forced to hire general labourers for repair work, who might not possess the specific skills required (Wilkinson and Chang-Richards 2016). In 2018, the government commissioned a public enquiry into the Earthquake Commission's handling of claims from the Canterbury earthquakes. The final report identified major failings in the Earthquake Commission practices, noting that 'shortcomings in its performance resulted in unacceptable stress, distress and delays in some people's recovery and repair of their homes' (Public Inquiry 2020, 8). At the same time, and while not excusing its failings, the report included some recognition of the difficult post-disaster context in which the organisation was forced to operate. This point can be illustrated with a case study of the coastal suburb of Southshore.

The case of Southshore

The intersection of land zoning and insurance settlement processes was a significant determinant of the recovery trajectory both of individual households – which might be insured with a different company than their immediate neighbours – and of entire suburbs across the city. The suburb of Southshore, a lower socio-economic community based along a narrow coastal spit in the east of the city, was particularly affected by these overlapping processes. The coastal spit, which runs almost north-south in orientation, is bordered by water on both sides, with an extensive river estuary on its western edge and the Pacific Ocean on the eastern side (Figure 1.1).

With approximately 600 households, Southshore was distinctive in that its properties were assigned to no less than three land zone categories: red, TC3 and TC2. The zones took the form of strips that ran longitudinally down the coastal spit, effectively dividing the suburb into three spatially demarcated household groups: those expected to leave (red-zoned), those where the land required further geotechnical investigations before any repairs could be undertaken (TC3), and those permitted to stay and repair their homes without any further investigations into their land (TC2). Although precise numbers were not publicly available, there were likely 200–250 households in each of these categories. In research undertaken in 2012 and 2013, a non-random sample of 20 households from each of these three zones completed a structured survey followed by a semi-structured interview. Although indicative rather than representative, this work offered a window onto the experiences, thoughts and feelings of Southshore residents regarding the zoning process (Conradson 2012; Dickinson 2013).

The government's land zoning announcements, as noted earlier, proceeded in stages rather than all at once for the entire city. The incremental nature of these announcements was of particular significance for Southshore. The first locally specific announcement occurred on 29 October 2011, when around 250 Southshore households were designated as being on green TC2 land. Most of the residents interviewed in this zone expressed a sense of relief that they were able to stay in Southshore, and that they could move forward with their repairs and rebuilding. Some also experienced a sense of guilt or reluctance to enjoy this outcome, however, as there were other households in the community still facing uncertainty while they awaited a zoning decision.

As one resident put it, 'I'm absolutely thrilled. But I can't celebrate until there's better news for everyone in the area'. As TC2 was the first land zoning category announced in Southshore, this group of residents had the shortest wait and the longest period in which to formulate their 'next steps'. The reduction in uncertainty was conducive to improved mental well-being, as found in other areas of the city (Greaves et al. 2015; Morgan et al. 2015). That said, a few of the TC2 households interviewed were sufficiently perturbed by their experience of the earthquakes that they had decided to move outside of Christchurch altogether, to places they considered to be more seismically stable and thus less likely to be disrupted. As 'green zoners' they had the opportunity to implement such intentions, in stark contrast to their subsequently designated TC3 neighbours (where the need for further investigations prevented any house sales that would enable quick departure) or red zoners (where leaving was essentially an involuntary matter, irrespective of one's preference for leaving or staying) (Conradson 2012).

At the time of the local green zone announcement (i.e. 29 October 2011), the rest of Southshore was placed into a 'to be determined' category. These households were then subject to several months of waiting, as well as more than one extension to the initially promised timeframes for a decision. In a letter to 401 households of 'west Southshore' in March 2012, the Minister for Earthquake Recovery, Gerry Brownlee, described the zoning determinations for the area as among 'the most difficult and challenging' in the city, given the complexity of the land damage it had experienced (Mann 2012). As the land damage varied along the western edge of the coastal spit, any uniform zoning categorisation would inevitably do some injustice to that on-the-ground variability. At the same time, a highly variegated zoning assessment with differentiated outcomes for individual and often adjacent properties would likely have provoked significant community dissatisfaction, potentially unleashing a deluge of correspondence and litigation.

In May 2012, the government announced it would red zone around 200 Southshore properties, all on the suburb's western estuarine edge. As well as the land damage, this decision took into account the subsidence in the southern part of the estuary, in the order of 0.3 to 0.5 metres (Environment Canterbury 2011). Among the affected households, some residents spoke of the psychological relief of 'going red', as it enabled them to move from a prolonged state of 'not knowing' to a position of increased clarity and freedom to act. One man explained that he was 'delighted to be in the red. I can't be doing with all this indecision', while a woman cheerfully noted that 'we broke open the champagne!' At the same time, several red-zoned interviewees recognised that leaving Southshore would disrupt their local friendships and connections, many of which were longstanding. The involuntary and disorienting nature of being red-zoned was often remarked upon:

> I am not happy at all. I have no idea what to do. Life is upside down. All my friends can stay, but we have to go. It is a strange story. We want to stay; they want to go. … I don't want to leave.

Still others expressed confusion about the red zoning decision, particularly when it did not appear to correlate with the housing damage they could observe:

> I have no idea what it is based on. If you go five metres down the road, they have turned green [ie. the property had been green zoned]. It doesn't make sense to me. My house is not that badly damaged.

Such a response highlights the challenging nature of accepting the outcomes of a zoning process that was based in part on an assessment of the likelihood of *future* land damage in the event of further earthquakes. The currently visible damage to a resident's house might well not be a good indication of this future liability, and this was understandably perplexing for some. In a few cases, homeowners who shared the same driveway received different land zone decisions, intensifying such confusion (Conradson 2012; Dickinson 2013).

Around 200 Southshore households were zoned TC3, and these residents experienced particular challenges. Although their land was considered habitable, it was also vulnerable to future quake-induced liquefaction and lateral spreading. Any substantial house repairs or rebuilds therefore required the construction of special foundations, individually designed for each house to compensate for the particular characteristics of its parcel of land. As demand for geotechnical expertise across the city significantly exceeded the supply of qualified engineers, the geotechnical investigations needed to determine the required foundations often took many months, and this slowed the process of household recovery on TC3 land (Greenhill 2013). In fact, many TC3 residents in Southshore complained of having no clear timeframes for the completion of their geotechnical investigations or repairs, which contributed to a sense of being 'in limbo'. Some were envious of the clear settlement timeframes their red-zoned neighbours had received, while others were uncertain whether their homes would be insurable in the future (Conradson 2012).

For people living in Southshore, the land zoning process thus had a major impact on their well-being and life trajectories. The government decisions compounded the material disruption from the quakes. As one Southshore resident explained:

> The earthquake was only the first catastrophe. What happened afterwards … the way it was handled, the way information was passed down, the inability of government to communicate with the people … yeah … that's another disaster in itself. People come up to me, you know, and say 'You're from Southshore? Poor you, you would have felt the earthquakes in a big way out there' … and I'm thinking to myself 'F*** the earthquakes, they were almost two years ago, but now I'm being kicked off my land and I don't have a bloody choice about it'. It's gut-wrenching.
>
> (in Dickinson 2013, 1)

Similar difficulties were evident across the city, given the extent of red zoning, and the amount of TC2 and TC3 land (Figure 3.2). In addition to the land zones, the post-quake experience of Southshore households was strongly influenced by their interactions with the Earthquake Commission and insurance companies (Miles 2016).

Psychosocial distress and mental well-being

In the immediate years after the earthquakes, levels of anxiety and depression rose among the Christchurch population as a whole (Beaglehole et al. 2019). These mental health problems were variously related to psychological trauma, involuntary relocation (which could disrupt long-standing attachments to place), grief in response to loss of life and injury, and the challenge of dealing with the Earthquake Commission and insurance companies (Morgan et al. 2015). The relatively frequent yet unpredictable aftershocks, illustrated in Figure 2.1, were also psychologically challenging, as they tended to keep pushing people into adrenalised threat states, while also undermining their perception of psychological security and environmental predictability. In September 2012, a survey was deployed to assess the subjective well-being of residents across the city. Eventually known as the Canterbury Wellbeing survey, this was administered twice yearly from September 2012 to September 2017, and annually thereafter. From the April 2013 survey onwards, a question on subjective well-being was included. Based around the WHO-5 well-being scale (Topp et al. 2015), this question made it possible to assess changes in self-reported well-being over time.

The sample size for each iteration of the survey was around 1200 individuals, drawn in a stratified manner to achieve representativeness with respect to gender, age distribution, ethnicity and household income. When the sample was divided into quartiles based on annual household income, there were clear differences in self-reported well-being between the four categories (under NZ$30,000; between $NZ30,001–60,000; between $NZ60,001–$100,000; over $NZ100,000) (Begg et al. 2021). The lowest household income group had the lowest self-reported well-being (and so on, through the spectrum, to the highest household income level, which had the highest self-reported well-being). This pattern was thought likely to reflect the well-observed positive relationship between income and self-reported well-being, which holds to a certain point. Although no pre-quake baseline measure for well-being was available, each income group began with a below-average level of well-being in 2013 which then rose over the following five years. This increase arguably reflected a process in which people were able to come to terms with the disaster and to return to a more steady or 'normal' form of life (Begg et al. 2021). As with any group data, such average figures obscure a host of individual and household variability, and there will be some in each group whose well-being did not substantively improve through the 2012–17 observation period (Begg et al. 2021).

Table 3.1 Issues having a moderate or major negative impact on Christchurch residents' well-being

September 2012 (top five issues, ranked by proportion of participants citing them)	September 2013 (top five issues, ranked by proportion of participants citing them)
1. Distress regarding ongoing aftershocks (42%)	1. Dealing with EQC/insurance issues in relation to personal property and house (23%)
2. Dealing with EQC/insurance issues in relation to personal property and house (37%)	2. Making decisions about house damage, repairs and relocation (21%)
3. Loss of other recreational, cultural and leisure time facilities (34%)	3. Being in a damaged environment and/or surrounded by construction work (20%)
4. Being in a damaged environment and/or surrounded by construction work (30%)	4. Loss of other recreational, cultural and leisure time facilities (17%)
5. Making decisions about house damage, repairs and relocation (29%)	5. Distress regarding ongoing aftershocks (12%)

Source: Morgan et al. (2015).

The surveys also asked participants to identify issues that had either a moderate or major negative impact on their well-being. Between September 2012 and September 2013, a small set of issues were identified as the most significant concerns by respondents in this regard (Table 3.1). In 2012, the most impactful issue for well-being related to the earthquake sequence itself ('distress regarding ongoing aftershocks'). This reflects the nature of earthquakes, in that they are typically accompanied by hundreds and even thousands of aftershocks (although as Chapter 2 has shown, in Canterbury, rather than occurring in a gradually attenuating fashion, the sequence was more drawn out). By 2013, the significance of aftershocks to respondents had fallen – from first to fifth ranked position – and the most negatively impactful issue had become dealing with the Earthquake Commission and private insurance companies. Two and a half years after the February 2011 earthquake, engaging with insurers was generating more distress for Christchurch residents than decisions about their property and housing situations, or the challenges of living in a damaged urban environment.

Unsurprisingly, exposure to earthquake damage was associated with increased levels of depression and anxiety. After the September 2010 quake, the residents of Avonside, an eastern suburb which experienced significant liquefaction, were found to have increased depression and anxiety scores (Dorahy and Kannis-Dymand 2012). After the February 2011 events, people across Christchurch experienced problems with hypervigilance, anxiety, irritability and difficulties sleeping (Potter et al. 2015). As well as the nerve-wracking effects of the aftershocks, these mental well-being impacts also reflect the inherent uncertainty of earthquakes – unlike a bushfire or flood, there is seldom any meaningful prior warning available – as well as their potential to rupture people's assumptions about a stable environment on which they can depend. After February 2011, there were short-term increases in prescriptions for anti-anxiety medications (Beaglehole et al. 2019). More generally, the stress of the earthquakes exacerbated pre-existing health problems for people, such as problem drinking (Thornley et al. 2013), and increased instances of employee burnout were observed, particularly when people's working conditions deteriorated after the quakes. This was due to issues such as job losses, increased volume of work, disrupted organisational processes and work environments, and prolonged strain in human service organisations (Thornley et al. 2013; Thornley et al. 2015; van Heugten 2014).

Particular groups experienced particular psychosocial impacts. Children who started school after the earthquakes were found to have heightened behaviour problems and post-traumatic symptoms, including difficulties with emotional regulation (Liberty et al. 2016). As Chapter 10 will show, these children were not helped by the government's post-earthquake decision to close many schools due to damage or population movements. In terms of demographic and social risk factors for quake-related mental distress, Hogg et al. (2014, 274) found that women and adults over the age of 39 'were more frequently diagnosed with mood and anxiety disorders than males and children'. New Zealand Europeans reported higher incidences of these disorders than the Pacific and Māori population in Christchurch, although Hogg et al. (2014) argue that this ethnic difference is likely to be an artefact of the tendency of Pacific and Māori populations to underutilise public mental health services, rather than a true reflection of their underlying mental health status in the aftermath of the quakes. In terms of geographic patterns, people living in eastern and central city suburbs, and in damaged areas of the Port Hills, showed an increased incidence of depression and anxiety, at rates around 23 per cent higher than the average for the Christchurch population (Hogg et al. 2014). They also had reduced subjective well-being (Greaves et al. 2015).

In February 2013, the Greater Christchurch Psychosocial Committee, a multi-agency consortium tasked with monitoring and promoting psychosocial well-being after the earthquakes, launched the *All Right?* mental health promotion campaign. The campaign used social media and conventional advertising to disseminate a range of mental health promotion messages (Figure 3.3). These often drew on phrases, humour and ideas sourced from the Christchurch community, and relayed them in a conversational rather than didactic register. A variety of posters were created for public settings – such as billboards, bus shelters, and public toilets – and for social media distribution. Since its formation, the campaign has continued as a collaboration between the Canterbury District Health Board (a regional body) and the Mental Health Foundation of New Zealand (a national body).

An evaluation found that awareness of the *All Right?* campaign among Christchurch residents had increased significantly since its launch – from around 51 per cent in 2013 to 92 per cent in 2020 – with similar visibility among both women and men, and the greatest awareness among younger people (Canterbury District Health Board 2021). Over four-fifths (86 per cent) of all respondents believed the campaign to be helpful for mental well-being and two-fifths (42 per cent) had engaged in one or more of the recommended activities. These were based on the Five Ways to Wellbeing initiative, with its focus on Connecting, Giving, Keeping Learning, Being Active, and Taking Notice (Mental Health Foundation, n.d). The examples in Figure 3.3 illustrate the emphasis on normalising a range of common post-disaster mental states and encouraging activities that are known to support mental well-being.

The increase in psychosocial distress and diagnosed mental health problems among the Christchurch population led the Canterbury District Health Board to request additional mental health funding from central government after the earthquakes. For several years, such requests were declined by the

Figure 3.3 Illustrative materials from the *All Right?* campaign.

Source: www.allright.org.nz

Ministry of Health, generating significant tensions in the relationship between the health board and the Ministry. This difficulty echoed other post-quake tensions between local and central government agencies, as discussed in Chapter 4. For some, the Ministry's refusal was further evidence of the poor treatment of Christchurch residents by central government agencies, and thus provided further energy for public protest and resistance by community groups (see Chapter 7). It was not until March 2016, over five years after the most damaging earthquake, that the National-led government made $20 million of additional funding available for mental health support services in the Canterbury region. This funding was eclipsed just a few years later, however, by the initiatives of the new Labour-led government which came to power in October 2017. In its 2019 budget, this government announced a record $NZ1.9 billion of additional funding for mental health services across the country. This investment has been directed to expanding early intervention initiatives such as the Mana Ake school-based programme (which seeks to help children learn skills to deal with loss, grief, peer pressure and bullying), as well as employing additional mental health professionals in community and acute services. Although there have been some problems with its timely uptake (Neilson 2021), this additional resource is gradually improving the support available to those in Christchurch who experience psychosocial distress as a result of having lived through the disruption and difficulties associated with the earthquakes.

Community and neighbourhood expressions of support

Alongside the initiatives of official agencies, novel forms of community and neighbourhood support arose following the earthquakes (Carlton and Vallance 2013). Residents provided food, water, clothing and temporary accommodation to others in their neighbourhoods; this was particularly prominent in the heavily disrupted eastern suburbs and lasted for several weeks, but was also evident to some degree city-wide. In general, people also engaged with their neighbours to a greater extent than usual. Many neighbourhood streets had increased pedestrian traffic and in-person interactions (as most central city workplaces were closed, with people working from home if possible, and many schools were also shut for a period of time). There were also a number of spontaneous creative gestures across the city, such as 'decorating fences with ribbons, flowers, and banners bearing messages of hope' (Thornley et al. 2013, 18), and placing flowers into the seemingly ubiquitous orange traffic cones that marked ongoing repairs to the city's roads and subsurface infrastructure. These small gestures complemented other quake-related creative expressions in the city, including street art, poetry and music (Feeney 2019; Fusté-Forné 2017; Norcliffe and Preston 2016). In a variety of ways, this creative work sought to offer people some encouragement and perspective during the often-depleting cycles of aftershocks, damage to homes and infrastructure, and negotiations with insurers.

These neighbourhood-level expressions of support often built upon existing social connections and relationships, as evident among interest groups, family, friendship and community connections. Whanau (extended family) and tribal structures were significant in the case of the Māori Earthquake Recovery Network. Formed in the aftermath of the February 2011 earthquake, this group made use of local marae (Māori meeting houses) as places of refuge and material support, and as spaces from which food, shelter and clothing could be accessed. It also drew upon whanau networks that extended across New Zealand. These networks facilitated people to temporarily relocate to other parts of the country, for rest and respite from the aftershocks, or because their homes had become uninhabitable (Kenney and Phibbs 2015; Potter et al. 2015). There was also a programme of residential visits in which Māori households, particularly in the eastern suburbs, were offered practical assistance with food, water, and information, with referrals made to other services as relevant (see Chapter 7).

In a variety of ways, these neighbourhood expressions of mutual aid and support assisted people to navigate what was, for many, a disturbing and challenging set of circumstances. Neighbourhood and community groups also helped to distribute information provided by government agencies, such as Civil Defence and CERA, to local residents and households (Paton et al. 2013). Those neighbourhood groups that 'mobilised quickly and pragmatically, using whatever resources they had available to address community needs' (Thornley et al. 2013, 18–19) typically supported community recovery. The facilities of some groups were repurposed as unofficial community hubs, including the Linwood Community Arts Centre and the Christchurch mosque on Deans Avenue, the latter of which became a hub for the local Muslim community. Some existing neighbourhood groups did not function effectively, however, for reasons that included ineffective leadership and internal conflicts, and this hampered their ability to make positive contributions to local recovery (Thornley et al. 2013). Alongside initiatives that focused on particular neighbourhoods, there were also a number of more broadly oriented responses, including from the Student Volunteer Army and advocacy and activist groups such as CanCERN, the Canterbury Communities' Earthquake Recovery Network (Chapter 7).

Conclusion

In the weeks immediately after 22 February 2011, the damage to people's homes and urban infrastructure, along with thousands of aftershocks, prompted substantial outmigration from Christchurch, both short and longer term. Although many people did return eventually, these departures were in any case offset by an influx of migrant workers drawn into the construction industry. Land zoning and insurance processes nevertheless affected thousands of households, and by 2013, when the frequency and intensity of aftershocks had attenuated, residents identified insurance matters as the issue having the greatest negative impact on their well-being

(Morgan et al. 2015). From 2013 through to 2019, however, self-reported well-being did show gradual increases for all income groups, which suggests a degree of recovery in the collective life and functioning of the city (Begg et al. 2021). Such an improvement should not be read as an easy confirmation of city-wide resilience, however, for in the same period there were significant increases in depression and anxiety, as well as multiple small business failures (Stevenson and Conradson 2017). In addition, it was not until 2016 – five years after the most damaging earthquake – that central government provided any significant additional funding to the Canterbury District Health Board for much needed mental health services.

Influenced by pre-existing patterns of inequality, the residents of particular suburbs and neighbourhoods in Christchurch often experienced quite different post-earthquake recovery trajectories (Wilson 2013). For households in the eastern suburbs or in lower-income groups, the quakes and difficulties with insurance claims saw significant declines in self-reported well-being. In contrast, residents in higher-income parts of the city fared comparatively better on average; the majority were able to resume a pattern of life in the weeks and months following the quakes not dissimilar to what they had beforehand. Nonetheless, those in suburbs like Fendalton whose homes bordered streams often suffered extensive housing damage, as did those on the hills, in places like Cashmere and Mt Pleasant. Such variations in post-disaster experience, socially and geographically, are of course typical of many disasters and not particular to the Canterbury earthquakes. They would nevertheless become one of the central issues in the city's recovery.

Part II
Recovery and renewal

4 Governance and the cartographies of recovery

Introduction

In a film made a decade after the Canterbury earthquake sequence, with the title 'When A City Rises', a prominent urban designer muses about whether the place that was emerging could have been 'an amazing beacon for how a new urban way of living can be created' (Lunday 2020). After ten years in which much of the city had been destroyed, demolished, re-planned and re-built, this streak of idealism was a reminder of the contests that had been fought about how the old and, in many respects, obsolescent city centre and its suburbs could be re-imagined for the future. Many of the debates centred as much on questions of process as much as outcomes. In other words, often the crucial issues in the Christchurch rebuild revolved around matters of governance, of who was to be responsible, how decisions were to be made, whose interests were included and in what ways.

This chapter is about the maps and plans used to reconstruct the city: mapping has long been one of the key instruments of governance employed by the modern state (Christensen 2013; Scott 1998). However, governance in the post-earthquake context is taken to mean more than just the activities of central and local government and includes the interests of business, of different communities and of the Treaty partner. Such a discussion helps to locate some of the experiments in governance that took place in the decade preceding the earthquakes, which had introduced collaborative and community-informed processes to map out possible futures for the city and its region. This perspective provides a basis for understanding reaction to the state-driven recovery mechanisms that were put in place in 2011, and the panoptical cartographies employed as part of re-settling the central city in particular. The extent to which the rupturing with the past (Pickles 2016) has led to loss of heritage and sense of place, or enabled a city to re-emerge in more engaging and inclusive ways, is then considered.

Issues of governance

The scale of damage wrought by the earthquakes and their impacts, outlined in Chapters 2 and 3, made it inevitable in retrospect that, after 22

DOI: 10.4324/9780429275562-6

February 2011, central government would become a key player in debates and delivery about the future of the city. A senior Christchurch planner described the Canterbury Earthquake Recovery Act of 2011, which enabled the introduction of large-scale, centralised planning into post-earthquake reconstruction, as 'the most powerful and draconian piece of legislation' that New Zealand had seen outside of wartime (Nicholson 2020). The reference to wartime is pertinent, as the trend internationally in the wake of the Second World War was 'to look to the war-time experience of mass production and planning as means to launch upon a vast programme of reconstruction and reorganization' (Harvey 1989a, 68; Judt 2010). The modernist style of planning and urban renewal that dominated the decades from the 1950s to 1970s was then generally accepted as a necessity for the delivery of social and economic goals. But by the new century, it had long since lost ideological traction amongst the Anglo-American polities in which New Zealand is generally situated. It did not however disappear completely, nor was it ever as coordinated, clear cut or unproblematic in practice as sometimes assumed (Beauregard 2020).

The neoliberal assault on the prominence and roles of the state, a process encapsulated in the phrase 'the New Zealand experiment', began here in the 1980s (Le Heron and Pawson 1996). A key feature of neoliberalism was the rejection of the egalitarian social democracy of the post-war years, with the Keynesian belief in the risks of market failure yielding to one of imagined government failure through inefficiency and lack of flexibility. The neoliberal shift involved 'the development of governing styles in which boundaries between and within public and private sectors become blurred' (Stoker 1998, 17). By the 1990s, the term 'governance' had therefore come to signal the engagement of a much wider range of players in the sphere of public action than previously. A feature of this shift was that good governance at the city level was 'now largely defined by the ability of formal government to assist, collaborate with, or function like the corporate community' (Hackworth 2007, 10). In Harvey's (1989b) characterisation, this was represented by a movement from managerialism to urban entrepreneurialism.

Harvey's argument was that this transformation of urban governance flowed from the recognition that, increasingly, the capacity to get things done rests not only with government, but also draws on wider sets of interests that can be more responsive to processes of competition between cities for production facilities, consumption flows and control functions. At the local level, one of the foundations of neoliberal governance has therefore been public-private sector cooperation. The Christchurch City Council, which after local government reform in 1989 became responsible for most of the metropolitan built-up urban area, worked to construct alliances with business interests to increase the attractiveness and tackle obsolescence of the central city in the face of heightened competition from suburban malls. It encouraged investment indirectly and directly in facilities such as a luxury hotel and mid-sized convention centre adjacent to the Town Hall (with its large auditorium and meeting spaces), also establishing an events company to service the complex. It also had some success in the decade before the earthquakes in facilitating development of old inner city lanes for retailing and nightlife (Johnston 2014).

The forms of governance that have emerged since the 1990s have however been more complex and elusive than simple categorisations like business-local government alliances imply (Arnouts et al. 2012). It is a much broader concept describing the constant renegotiation of national and local state relations within both market and society. The term 'collaborative governance' has been adopted to denote processes that develop interdependencies amongst a range of actors, whilst seeking to establish political legitimacy for public actions. Such processes have been characterised as formal and deliberative, initiated by public agencies to include non-state actors (Ansell and Gash 2008). An example was Canterbury Dialogues, 'an experiment in civic participation' held over eight months in 1998. It included parties from local government, business, community groups and academia, hosted by the Canterbury Employers' Chamber of Commerce. It focused on 'managing urban growth, developing sustainable economic wellbeing, effective environmental management, and social cohesion' (Wagner 1998). A similar mix of parties participated in the Future Path Canterbury visioning process coordinated by the regional council, Environment Canterbury, in 2001 (*City Scene* 2001).

These initiatives provided local experience in drawing on wider ranges of expertise, lending confidence to more collaborative forms of governance. A formal version emerged in 2004 when Environment Canterbury, the three local authorities in the wider Christchurch urban area (Christchurch City, Selwyn and Waimakariri District Councils) and Transit New Zealand (central government's transport infrastructure provider) set up the Greater Christchurch Partnership. The outcome was the Greater Christchurch Urban Development Strategy (2007). It established a framework for accommodating new metropolitan housing developments through means of both intensification and greenfield expansion, whilst protecting significant resources, such as water catchment areas. The framework was the outcome of public consultation on a range of options and was declared to be the basis for the organisations and community to cooperatively manage urban growth. The Strategy became a critical document in post-earthquake recovery, but arguably in a compromised fashion, as is discussed in Chapter 5. Figure 4.1 illustrates the timing of this and other key institutional reforms.

The partnership did not formally involve Ngāi Tahu, whose takiwā (tribal territory) includes the Canterbury region. But when the Greater Christchurch Urban Development Strategy (2016) was updated, Ngāi Tahu was included as one of the partners. This illustrates that just as actually existing forms of neoliberalism are contingent on context (Brenner and Theodore 2002), so too are modes and shifts in governance. Since the 1980s, specific pieces of national legislation have required that the principles of the Treaty of Waitangi, signed in 1840 between Māori tribes and the Crown, be recognised in governance processes (Finlayson and Christmas 2021). How to give effect to this in particular contexts has been a matter of acknowledgement

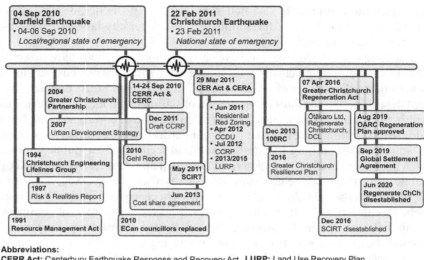

Figure 4.1 Timeline with key institutional reforms.

Source: Adapted from Huck et al. (2020, 5)

and negotiation over time. This has also been the case with resource management where the interests of multiple stakeholders, including more-than-human actors, are involved (Peart and Cox 2019). The recognition of Māori cultural perspectives and property rights, as well as the dependence of both market and society on the more-than-human realm, are illustrations of the context-specific shape of what is increasingly referred to as 'network governance'. Such recognition would also appear to represent steps towards acknowledgement of intergenerational interests.

A definition of network governance is 'a relatively stable cooperative arrangement between independent actors based on trust and reciprocity' (Bednar et al. 2019, 703). But network governance has been described as only working effectively if the networks function well (Lewis 2011), and critiqued for shortcomings in achieving material outcomes due to the dispersed, non-hierarchical nature of the network. In the case of wicked problems such as climate change, it has been argued that it 'must be supplemented with other governance modes to achieve effective implementation' (Bednar et al. 2019, 714). On the other hand, it is vulnerable to co-option by powerful parties in the network (Cretney 2019). A clear example of this in Canterbury was the decision of a centre right National-led government in 2010 to legislate to remove the elected members of the regional authority, Environment Canterbury, and impose its own commissioners, in part to facilitate the needs of an expanding dairy industry. This conflicted with the responsibilities of the elected represent-atives which, shaped by the Resource Management Act, were more broadly based and longer term (Mueller 2017). The decision also reflected what has been described as 'a systematic governance problem', a tendency in multiple policy domains to favour short-term interests, or a 'presentist bias' (Boston 2017, 15).

Even before the introduction of the Canterbury Earthquake Recovery Act in 2011, there was therefore a vivid example in the region of the hierarchical imposition of power overriding democratically sourced local interests. But there had also been, in the decade or so before the earthquakes, developing experience in the city and urban region of Christchurch with collaborative and networked forms of governance. It would be a mistake however to portray this in starkly oppositional terms as a conflict between central power and local interests. The reality is both more nuanced and engaged, with collaborative experience and initiatives informing and, to an extent, shaping, central govern-ment's post-earthquake response. But as the subtitle and content of the next section indicate, this is often not how it was publicly heard or seen in the city.

Panoptical maps and plans

A landmark event in the early earthquake years that still resonates in public memory is a city council-sponsored forum called 'Share an Idea' (Bennett et al. 2014a). This was organised with the assistance of Jan Gehl's interna-tional urban design consultancy. Gehl, from Denmark, had been engaged by the council before the earthquakes, to identify ways in which the central city was sub-optimal (Gehl Architects 2010). The site is flat and was developed within a relentless colonial grid, in which low-rise older masonry buildings predominated, with fewer of the tall, modern buildings that had come to char-acterise the North Island centres of Wellington and Auckland. The signifi-cance of Gehl's report from a governance perspective is that he brought to Christchurch a reputation for retrieving traffic-dominated cities for pedestri-ans and street life, based on close observation of how people use public space and working with citizens in participatory ways. His report reflected earlier Gehl projects in cities such as Copenhagen, Barcelona and Melbourne, rec-ommending a more pedestrian- and cycle-friendly city, with a 'traffic calmed' centre, inviting and accessible public spaces, and 'green blue' connections that emphasise amenity features such as parkland and the Ōtākaro Avon river.

The city council was therefore well-placed when given responsibility under the Canterbury Earthquake Recovery Act 2011 to draw up a plan for city centre renewal. Gehl Architects returned to assist in the public consultation and planning process that took place in mid-2011. Share an Idea ran over six weeks, beginning with a two-day expo attended by 10,000 people, and includ-ing public lectures from local and international speakers, mail drops, stake-holder meetings and public workshops. Over 106,000 ideas emerged from

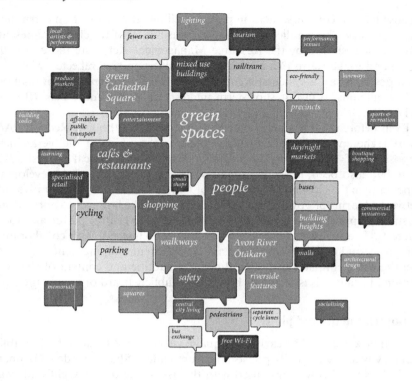

Figure 4.2 Themes emerging from Share an Idea.

Source: Adapted from Central City Recovery Plan (2012, 22).

'the highest level of community involvement ever seen in New Zealand' (Nicholson and Wykes 2012, 31). These are illustrated in Figure 4.2; many were in tune with Gehl's earlier report and provided the opportunity to act on its central theme of 'a change of mindset ... where traffic planning and public space planning are thought of as one' (Gehl Architects 2010, 117). The resulting city council recovery plan was described in Mayor Bob Parker's foreword as 'the People's Plan for their city' (Christchurch City Council 2011, iii), reflecting a strong sense of local engagement. *The Press* portrayed it as 'winning widespread praise from residents' (Sachdeva 2012). It was an outcome in line with local expectations of being heard, based on a decade or more of collaborative governance experience. The reaction of the business community was however more critical, as Chapter 6 will detail.

Share an Idea won an international award from the Netherlands-based Co-Creation Association, the first time it had been given outside Europe. The chair of the jury remarked that 'Next to developing the new city centre, the result of this co-creation is also a stronger community' (Scoop 2011a). A key question however was how central government saw its role in the context of disaster, following the destruction of the South Island's main commercial centre and tourism hub. Initially, after the September earthquake, Parliament had passed the Canterbury Earthquake Response and Recovery Act 2010. This opened the door to intentionally hierarchical interventions, permitting government ministers to suspend or make exemptions to almost any New Zealand law with minimal restriction, in order to expedite the recovery process. As law, it was heavily criticised by legal academics and the New Zealand Law Society (Anderson et al. 2010; NZLS 2010). The powers were intended to be of fixed duration, with the Act expiring on 1 April 2012. However, long before then, it was superseded by the Canterbury Earthquake Recovery Act 2011, passed in the wake of the devastating event of 22 February that year (Figure 4.1).

The new Act, like its predecessor, was time-limited (for five years in this case) and did not specifically override the authority of the city council, although this body was by then regarded, including by councillors, as divided and ineffective (Vallance and Carlton 2015). The Act created a new government department, to be based in Christchurch: the Canterbury Earthquake Recovery Authority (known in the city as CERA, pronounced 'Sarah'). Such a move is one of 'the most visible governmental responses to disaster' internationally (Johnson and Olshansky 2017, 314). The position of Minister for Earthquake Recovery, with widespread powers, was continued from the 2010 Act. These powers enabled various pieces of existing legislation to be overridden, including those allowing for participatory and anticipatory forms of governance, such as the Resource Management Act 1991 and the Local Government Act 2002. In effect, the Canterbury Earthquake Recovery Act set up new structures 'without clearly establishing a joint governance model' with local elected councils, 'or clearly distinguishing between their respective roles' (Smith 2014, 147). It did however specify that Ngāi Tahu was to be a partner with the Crown in the rebuild. Further legitimacy for the overall approach was sought in the appointment of the head of the local power lines company (who was widely credited with restoring power quickly after each major quake) as Chief Executive of CERA, and in retaining as Minister a senior local Member of Parliament for the governing National Party, Gerry Brownlee.

The view from Wellington, or from CERA's offices in one of the few undamaged high-rise towers in central Christchurch, was inevitably more panoptical than the collective spirit expressed from the grassroots in Share an Idea and reflected in the 'People's Plan'. Ironically, central to Jan Gehl's urban design philosophy is the characterisation of good design as working from eye level and prioritising the relation between urban life and form. Gehl (2015) has contrasted this with the modernist approach to city building, where the architect models form at building height, and the planner envisages the totality higher still from a bird's eye perspective (cf. Scott 1998). He refers to this as the 'Brasilia syndrome', the utopian capital city of Brazil conceived in the 1950s as a visual whole when seen from the air, in which everyone is presumed to have a car and the pedestrian is an irrelevance. The plan for Brasilia was ruthlessly panned by the art critic Robert Hughes as 'an expensive and ugly testimony' to thinking 'in terms of abstract space rather than real place, of single rather than multiple meanings, and of political aspirations rather than of human needs...' (Hughes 1991, 211). No one has suggested that the post-earthquake redevelopment of Christchurch apes Brasilia, but a great deal of the reaction to it has reflected the humanist, street-scale perspectives of luminaries such as Gehl and Jane Jacobs (Jacobs 1961; Scott 1998).

Government's perspective on its top-down approach to post-earthquake reconstruction was that it 'would remove local confusion and provide the greatest certainty' (DPMC 2017, 7). In part, this was based on a review of responses to international disasters. The state could also resource relatively intense land assessments throughout residential areas for future suitability for settlement in the face of environmental hazards. On this basis, CERA produced a detailed map identifying green and red zones, discussed in Chapter 3 (Figure 3.2). The green zones were subdivided into three categories according to the future risk of liquefaction, with requisite foundation systems for repaired or replacement housing in the corresponding areas (Saunders and Becker 2015, 78). Red zones included the Avon Ōtākaro river corridor that had been badly damaged due to liquefaction and lateral spread, in which the risk of further damage, not least from flooding, was too great and the success of remediation, let alone its affordability, was in question. Smaller areas of red zone due to the ongoing risk of rockfall were declared in the hill suburbs, as well as alongside the river in the satellite town of Kaiapoi. To resolve these matters quickly, residents received a Crown payout for forfeiting their red zone properties through the extraordinary powers vested in the Minister (Smith 2014). The consequences of this policy are explored in Chapter 10.

In the central city, however, the 'People's Plan' was controversially overturned within a few months (Cretney 2019). In April 2012, Minister Brownlee – anxious to counter any flight of business investment – directed CERA to set up a Central City Development Unit. This unit was to oversee the preparation of a more detailed 'blueprint' for the central city in 100 days. It reflected central government's belief that the council plan, self-costed at $2 billion, did not sufficiently allow for the 'focused, timely, and expedited recovery' specified in the Canterbury Earthquake Recovery Act (DPMC 2017). A new Crown blueprint team was selected quickly. It was a consortium of six firms, including a planning consultancy and two architectural practices from Christchurch, alongside an Auckland project manager and two Australian urban planning companies. The team divided itself into two groups. One refined a broad vision, building on and incorporating many of the elements of the council's plan. The other identified and located a series of ambitious 'anchor projects', to be funded by negotiation between central and local government (McCrone 2012a). These were intended to retain business confidence and provide a framework of 'precincts' on the ground within which private investment would flourish, and the state's investment be protected. There was to be little more public engagement (Amore et al. 2017).

The differences between the blueprint – officially the Central City Recovery Plan (2012) – and the council plan (Christchurch City Council 2011) were therefore both of product and process. The blueprint was launched in completed form on 30 July 2012 and came into law the next day. In contrast, the People's Plan had been subject to feedback from a travelling roadshow, and a public submissions process as provided for in the Local Government Act 2002. It was less detailed than the Crown blueprint, but embodied the spirit of Share an Idea and was clearly intended to allow for a relatively organic and granular of rebuilding. At the time, post-earthquake Christchurch was already becoming known internationally for many innovative transitional activities (Bennett et al. 2012), as outlined in Chapter 8. Both plans envisaged a more focused, low-rise city centre, but the blueprint adopted more visibly many of the tenets of modernist city planning. It was a sweeping cartographic vision, dependent on widespread Crown acquisition of land, and the amalgamation of titles to form large development plots. Within these, the anchor projects and surrounding themed precincts (for 'innovation', health, cultural activities, sport, retailing, etc) were to be planted within or across city blocks. These would be ringed by 'the frame', defining a central area 'of a scale appropriate to current demand' (Central City Recovery Plan 2012, 33). In its ambition, rationality and efficiency, the blueprint, and the map that embodied it, could be seen as the epitome of what James Scott (1998) described as thinking like a state (Figure 4.3).

Business activation alliances

The blueprint not surprisingly became performatively powerful, but not to the extent of sweeping all before it. The assumption of centralised authority did not silence other players; indeed it was intended to provide the framework for a range of entrepreneurial initiatives forged between the business community, central and local government (discussed in Chapter 6). Some of these initiatives bear directly on the theme of governance. The most immediate was an Earthquake Support Subsidy agreed after the Minister of Finance approached the Chamber of Commerce to negotiate how to maintain business cashflow (Roughan 2017). Initially, from September 2010, this was targeted to small businesses, then opened up those of any size in February 2011. It recognised the problems that up to 6000 businesses in the central city faced. It was complemented by actions by banks (advancing money), Inland Revenue (forgiving or delaying payment of goods and services tax) and insurance companies (providing rapid partial payouts). These were high trust schemes designed to stem business failures and worked so well that business churn through the earthquakes was no higher than the usual turn rate of about 11 per cent each year (Townsend 2016). The Earthquake Support Subsidy also provided the template for the Labour-led Government's response to business and employment needs in the Covid-19 emergency ten years later.

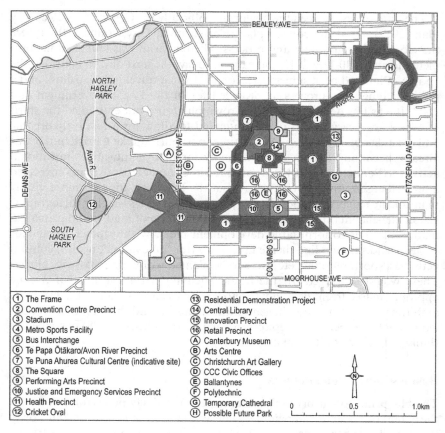

Figure 4.3 The blueprint map.

Source: Redrawn from Central City Recovery Plan (2012, 33–4).

Many affected businesses relocated to the largely unscathed western and southwestern parts of the city. Some displaced technology companies moved to the Enterprise Precinct and Innovation Campus (EPIC), one of the first commercial developments in the central city, which opened in late 2012. Its campus-like building is also designed to accommodate business start-ups and networking events (Andersen and McLennan 2014). For some years it was the home of the Ministry of Awesome, also founded in 2012 to provide advisory support and mentorship (Anderson 2012). The establishment of EPIC was supported by CERA, the city council and government agencies, with founding partners including the Bank of New Zealand and Google. It is also self-described as the 'foundation partner' in the blueprint's Innovation Precinct (EPIC 2022), and as such has outlasted the other rapidly established post-earthquake initiative in the central city, the Christchurch container or RE:START mall.

RE:START was established by a trust set up by inner-city retailers, in part of what the blueprint subsequently designated the retail precinct. It opened at the end of October 2011 when most of the cordon – which was not finally removed until 30 June 2013 – was still in place. RE:START was located adjacent to (and in conjunction with) Ballantynes, an undamaged department store, with assistance from CERA and the city council (Bennett et al. 2012). It quickly became not only a critical initiative in drawing shoppers back to the central city but also in building the city's reputation for urban transitional activities in the post-earthquake years. In 2013, *Lonely Planet* listed Christchurch amongst its top ten cities internationally, highlighting the colourful labyrinth of containers in the mall (Atkinson et al. 2013). It was originally to stay until 2014, but after relocating twice along the street to make way for new retail and office buildings, did not finally close until 2018. Many of its tenants were rehoused within new developments, discussed in Chapter 9.

RE:START was a conspicuous success; less conspicuous but nonetheless foundational was the Stronger Christchurch Infrastructure Rebuild Team (SCIRT), which began work in May 2011. SCIRT has attracted some attention in the business literature as a collaborative project-based alliance that brought together five of the country's largest contracting companies, funded by CERA, the city council and Transit New Zealand. It was tasked with the delivery of about 600 projects to repair the city's horizontal infrastructure (Walker et al. 2017). SCIRT created its own structure, workflow and processes amongst a group of usually competing contractors with similar areas of capability: a model designed to operate in a situation of urgency, resource constraints and public expectations that water supply, sewerage and road surfaces would be serviceable again within reasonable timeframes. Although SCIRT's work was compromised by political contests over the extent of repairs and of the funding of them (Huck et al. 2020), by the time it was wound up in late 2016, it had substantially achieved its goals. Remaining repairs were incorporated into city council's ongoing asset management programme.

Heritages lost and found

Business activation alliances were therefore essential in supporting repair of the city's infrastructure, and in maintaining and to an extent developing its employment base. In the central city, these alliances were important enablers of the blueprint which framed the direction of the central city rebuild within a complex context of demolition and renewal. Demolition was the responsibility of the Central City Development Unit, which was authorised by the Minister for Earthquake Recovery to issue 'section 38' notices to override any public say in the removal of buildings, many of which were inaccessible behind the civil defence cordon (illustrated in Figure 6.1). The intention of the Central City Recovery Plan (the blueprint) was that the process of reconstruction would be led by the state's investment in the rebuild, particularly the anchor projects. These were to be financed through a controversial cost-sharing agreement between central government and the city council, developed with little consultation and signed in June 2013 (Johnson and Olshansky 2017). The agreement codified the extent of the Crown's commitment to reconstructing central Christchurch on a project-by-project basis. If it was substantial in financial terms, the price was the forfeit of local control, the imposition – as many saw it – of outsized projects and the loss of a great deal of the city's 'heritage' in the name of the map.

In a discursive critique, Farrell (2015, 19) avers that the blueprint was 'just the latest in the long catalogue of maps that have attempted to draw this country, to reduce its bewildering complexity to something comprehensible'. In Christchurch, the first of these was the original colonial settlement grid superimposed across the sand dunes and wetlands of the site chosen for the city in 1850. Its indiscriminate geometry came at the expense of the destruction of indigenous ecologies and disguise of the land's liveliness, as earlier chapters have shown. The blueprint did little to disturb the original grid although, like the council's plan, it placed more emphasis on the river corridor as a transgressive feature. It also carved out the well-defined 'frame' to contain the central city and provide active transport corridors and space for downtown housing (Figure 4.3). In this, it echoed the town belts of the 1850 plan. The extent of demolition within the blueprint did however remove many of the streetscapes and landmarks that had accreted across the grid over generations, and with it any pre-earthquake sense of place. To Farrell (2015, 103), 'In this city, it is easy to feel lost'. This is the context for often vigorous post-earthquake debates about heritage.

As an international phenomenon, Lowenthal (2004) dates what he calls 'the heritage crusade' from about 1980. Four years later *Lost Christchurch*, a book of buildings demolished over time to make way for new developments, was published. Lacking a dramatic site or 'individually outstanding' buildings, the author observed that what gave the city 'an individual character is the juxtaposition and association of different buildings and the relative scale of those buildings' (Wilson 1984, 11). When during and after the earthquakes, hundreds of buildings were demolished in the central city, including about 250 on the city council's own heritage list (Blundell 2021), a reaction was inevitable: heritage is most valued when most at risk. Heritage advocates disputed the Central City Development Unit's indiscriminate use of section 38 notices, based on whether or not buildings were repairable. A lot were sacrificed to the blueprint map, as standing in the way of anchor projects or precinct developments. But often demolition was at the request of owners who lacked desire or resources to undertake repairs (Gray 2016), which in turn contributed to the visual dominance of unmade car parks for many years.

Some traditional architectural heritage has survived in key locations, such as the Gothic Revival Canterbury Museum (which had been seismically strengthened), and much of the Arts Centre opposite it, a complex of buildings in the same style that until the 1970s had housed the University of Canterbury before its suburban re-location. The Arts Centre was badly damaged but has undergone a lengthy restoration programme, funded largely by insurance and employing up to 20 stonemasons from around the world at times (Erskine 2022). But the survival of such landmarks did little to lower the temperature in debates over the future of the city's two cathedrals: the Anglican Gothic Revival building in Cathedral Square, at the centre of the street grid, and the rather more stately Catholic basilica on the edge of the central city, sometimes held to be 'the finest Renaissance-style building' in Australasia (Ansley 2011, 81). The future of the former was subject to dispute for many years, after a church decision to demolish it, and civic and heritage campaigns in opposition. Eventually, a resolution to reinstate the building was brokered and supported in 2018 by the Labour-led government that had come to office the previous year (Bohan 2022). Restoration began a decade after the upper part of the tower and spire had collapsed into the Square. A decision to partially repair the wrecked basilica was overturned about the same time when a new Catholic bishop decided to build a new facility in the central city (see Chapter 11).

More public energy has perhaps been expended on the future of the cathedrals than any other issue in the post-earthquake city (Blundell 2021), demonstrating that for many the social value of heritage transcends the private interests of its owners. Cathedrals are historical landmarks, and the Cathedral in the Square was for long the stylised symbol of the city's branding (Galloway 2014). Nonetheless, such debates prompt consideration not only of what has been lost but also of opportunities opened for the expression of identities and interests not well served in the pre-earthquake city. Its colonial Gothic architecture and self-styled 'Garden City' image spoke to many of a certain provincialism. Yet – as observed in Chapter 1 – politically it is largely a left-leaning Labour town, out of tune with the centre right National-led government that was in power from 2008 to 2017, and often at odds with the Minister for Earthquake Recovery whose parliamentary seat was for many years the city's only 'blue' stronghold (and even that went to Labour in the 2020 parliamentary election). Mr Brownlee's active support for the programme of post-earthquake demolition symbolised this. Yet at the same time, the earthquakes presented an opening – with the crumbling of the city's material facades – to engage progressively with difference, in both human and more-than-human realms (Pickles 2016).

Parliament itself recognised this with the inclusion of Ngāi Tahu as a statutory partner in the Canterbury Earthquake Recovery Act. For Māori, this reflects the centrality of the principle of rangatiratanga, which broadly aligns leadership and authority (Pauling et al. 2014). Staff from Ngāi Tahu agencies worked with the city council on the People's Plan and on the Crown's blueprint. They established the Matapopore Charitable Trust, which is responsible for ensuring that tribal and sub-tribe (Ngāi Tahu and Ngāi Tūāhuriri) values, aspirations and narratives are realised within the recovery of the city. An important aspect of this is kaitiakitanga, which recognises the special relationship that tangata whenua, people of the land, have with local ecology. In this way, the exotic plantings of 'the Garden City' have been reworked throughout the implementation of the blueprint with an interweaving of indigenous and introduced plant species. An example is in Te Papa Ōtākaro, the Avon River precinct, which delivers the desire to reconnect with the river in the inner city, identified in both the Gehl report and Share an Idea. The design for this new two-kilometre-long pedestrian promenade includes a series of whāriki, or paved patterns resembling woven mats, which tell some of the Māori stories of place, whilst reflecting manaakitanga, the provision of hospitality and care to manuhiri, or visitors (Figure 8.1).

Te Papa Ōtākaro, in pedestrianising the river corridor in a city whose street grid had increasingly dominated by cars, is an anchor project that has become extremely popular. So too is a second anchor project, Tūranga, the new city library opened in 2018. Tūranga is the homeland of the Ngāi Tāhu ancestor, Paikea, and the name carries 'considerable responsibility', speaking of whakapapa, or affiliations across generations, and relationships across space. The joint Danish-New Zealand design was developed in conjunction with Ngāi Tahu and Ngāi Tūāhuriri, so that the building's terraces and openings on the upper floors face culturally significant points in the Canterbury landscape. Its golden aluminium façade reflects the folds of the Port Hills on the city's southern edge, as well as the shape of the leaves of the harakeke flax plants that originally grew on the site. Inside, the ground floor reflects the desire to welcome people, with the stairs and central atrium echoing the Ngāi Tahu narrative of knowledge (Matapopore, n.d.). The second level is largely for children and houses the Imagination Station, a creative play area run by a social enterprise that originated as a post-earthquake transitional initiative.

Children and their parents have also flocked to a third anchor project, the Margaret Mahy Playground, discussed in more detail in Chapter 9. Named after a much-loved New Zealand children's author, and located at the eastern end of Te Papa Ōtākaro, it was designed in collaboration with children. It is – not surprisingly given the ambition of the blueprint – the largest play area in the country. As a children's place, it is connected to active transport corridors in the frame and along the river, along which are interwoven the stories and symbols of tangata whenua (people of the land). Examples such as the playground, Tūranga and Te Papa Ōtākaro demonstrate that heritage is 'an ever-changing palimpsest', in which 'new creations and recognitions more than make up for what is lost through erosion, demolition, and changing tastes' (Lowenthal 2004, 39–40). In this sense, and despite reaction to the Crown's blueprint, the spirit of Share an Idea and Jan Gehl have prospered in the city.

From rebuild to regeneration

The five-year term of the Canterbury Earthquake Recovery Act expired in 2016. But how much of the blueprint had been delivered by then? What forms of governance replaced CERA and the Central City Development Unit, and with what sort of outcomes? In a retrospective collection, 'Christchurch – Five Years On' in *Architecture New Zealand* magazine, Peter Marshall, head of a prominent architecture practice and member of the blueprint team, noted that in the 13 city blocks that now comprised the compressed business district, 'building activity is under way in nine, including many of the anchor projects identified in the Blueprint'. Given the extent of clearance undertaken, he considered this 'an extraordinary achievement' (Marshall 2016, 38). But Hugh Nicholson, then the principal urban designer at the city council, and the lead on the 'People's Plan', observed that 'In our worst case scenarios, we never imagined' that so many buildings in the central city would be demolished, before noting that the replacement style is a kind of 'corporate urbanism' of commercial buildings with open floor plans. Banks were only financing new buildings with confirmed tenancies, and the only tenants with long enough lead-in times and deep enough pockets were 'the larger corporates' (Nicholson 2016, 47–8).

These observations warrant further context. The corporate style was also a product of the blueprint's requirements that developments be of a certain size, height (not over seven storeys) and strong enough to meet the revised building code. Combining these requirements with the pace of demolition, developers were for some time more active outside, and along the outer edge of, the blueprint, as Chapter 6 reveals. In a demonstration of the performative power of the map, their projects initially activated an axis connecting Victoria Street, a popular pre-earthquake shopping street, with Durham Street, which faces the western side of Te Papa Ōtākaro (Figure 6.4). This axis continued beyond Hagley Park into the inner suburb of Addington. In all three locations, high-quality business accommodation for law and accountancy firms was being built by 2012. Much of the commercial reconstruction coming on stream by 2016 within the blueprint area was not only for corporate clients, such as banks, but for lease to government departments. In the retail precinct, the RE:START mall, although relocated several times to make way for new buildings, still remained in place. The only completed anchor projects in the central city were the new Bus Interchange, and the Margaret Mahy Playground. Others were under construction, but some had not been started. The 'anchors' were certainly not leading the rebuild as originally intended (Figure 4.4).

In late 2014, the Minister for Earthquake Recovery established an advisory board to consider the transition away from Crown direction of the rebuild. It comprised leaders from local government, Ngāi Tahu, business and community groups. The message was a clear desire for greater local control. The outcome was the Greater Christchurch Regeneration Act 2016. The shift away from 'recovery' to 'regeneration' was intentional. Regeneration was defined in section 3 (2) of the Act as rebuilding, and 'improving the environmental, economic, social, and cultural well-being, and the resilience, of communities'. A key instrument in this regard was the establishment of a joint Crown-city council agency, Regenerate Christchurch. This agency was given authority to develop regeneration plans for different parts of the city. There was cross-party agreement in parliament that the time was right for such a move, and optimism that the new agency would carry on some of the work initiated by CERA, but enable this to be led 'by local people'. Although the new Act again had a lifetime of five years, Minister Brownlee (2016) expressed

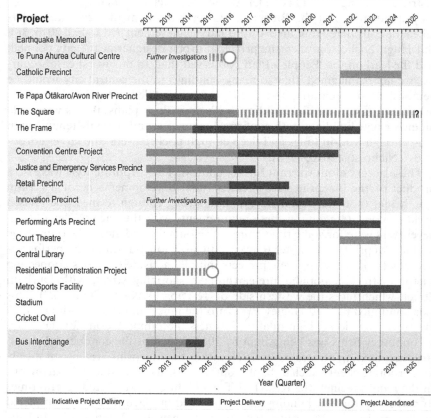

Figure 4.4 Planned and actual timelines of the anchor projects.

Source: Adapted and updated from Central City Recovery Plan (2012, 46)

the hope that 'over a period of years', Regenerate Christchurch would transition to a 'fully council-controlled organisation'.

Regenerate's main achievement was to be the production of a regeneration plan for the Ōtākaro Avon River Corridor (Regenerate Christchurch 2019), the new moniker for the riverside residential red zone created in 2012 (Figure 3.2). The plan, and the exhaustive public consultation on which it was based, is one of the main themes of Chapter 12. It received Ministerial approval in 2019. Other Regenerate initiatives, such as a development plan for the future of Cathedral Square (Regenerate Christchurch 2018a), have not been progressed. The Square, with the Anglican cathedral at its centre, was originally intended as the focal point of the central city. But it had struggled for many years and was subject to changes in design from the 1960s on. Since the earthquakes, the protracted debates over the repair of the Anglican cathedral, which is unlikely to be finished much before the end of the 2020s, and the shift towards the river with the success of Te Papa Ōtākaro, contributed to a lack of enthusiasm for Regenerate's Plan (Hayward and McDonald 2019).

Sometime before the expiry of the Greater Christchurch Regeneration Act, both government and city council had shifted away from their original view of Regenerate's potential as a collaborative partner in the rebuild. An official review completed in 2019 considered that a 'tipping point has been reached where the need for the legislation in its present form is effectively over' (Sinclair 2019, 5). The city council had regained powers to administer its own District Plan processes, following the revocation of a special Earthquake Order made in 2014. Regenerate's specified role for climate change planning on the New Brighton coastline was also transferred to the city council in 2019. The Act was amended in 2020 to terminate the agency's existence, with its functions being transferred to local agencies including the council. This followed a gradual loss of political confidence due to frustration with the number of rebuild agencies operating in the city. One agency did however survive. This was Ōtākaro Ltd, a Crown-owned company responsible under the 2016 Act for completing the anchor projects.

Ōtākaro completed Rauora Park, the central axis of the east frame, in late 2017 and Te Papa Ōtākaro a year later. But three largest anchor projects, discussed in detail in Chapter 9, were considerably delayed. Te Pae, the convention centre, opened at the end of 2021, at which point Parakiore, the metro sports facility and Te Kaha, the 'Canterbury Multi-Use Arena', (popularly known as 'the stadium'), were still anticipated (Figure 4.4). The delays have been due to site issues, contracting capacity and available finance, but cast the original intent of these blueprint projects as 'anchoring' private sector developments in a different light. The scale of the blueprint's ambition required additional Crown funding for the stadium, and long-term ownership of Te Pae, after the city council refused to take responsibility for this behemoth in the Global Settlement Agreement in late 2019. The global settlement was negotiated between the city council and the country's then Labour-led government. It determined the future ownership of Crown and council regeneration assets and was a milestone in the withdrawal of the Crown from post-earthquake governance in the city (Global Settlement Agreement 2019).

Conclusion

To what extent, returning to James Lunday, has the post-earthquake rebuild of Christchurch been 'an amazing beacon for how a new urban way of living can be created'? In the face of such a complex, long drawn-out process, views and assessments inevitably differ. The architect Peter Marshall, a member of the blueprint team, saw 'something unique' happening in a place 'redefining itself as an Arcadian city in the South Pacific' (Marshall 2016, 39). A few years later, John Walsh of the New Zealand Institute of Architects quietly asserted that 'the procrustean planning regime – Modernism redux, with zones recast as "precincts" – imposed by the planning authority foreclosed the possibility of new paradigms of urban development' (Walsh 2020, 159). Both seem like easy conclusions to reach. But as this analysis of the rebuild process and outcomes has shown, the reality has been more complex, as experiments in governance, mapping and regeneration, and heritages revealed as well as lost, indicate. The following chapters may show in greater depth whether or not 'a new way of urban living' is a chimera.

5 Housing recovery

Introduction

Most of the housing stock in Christchurch city and adjacent areas of its two neighbouring local authority jurisdictions, Selwyn and Waimakariri districts, which together comprise the Christchurch urban region, was in some way affected by the earthquakes. This disrupted lives and created a complex and protracted response from house owners, renters, the insurance sector, central and local government politicians and planners, residential property developers, their financiers and the building industry. The objective of this chapter is to outline how these participants in the recovery process interacted as the subsequent housing regeneration effort unfolded. It will show how this recovery was linked to transport infrastructure development after the earthquakes and produced an array of outcomes in the central city, suburbs and satellite towns. The recovery has intensified the central city and parts of the suburbs residentially, but in an illustration of path dependency (Dodds 2021), also significantly increased a long-established tendency to urban sprawl on the margins of the region (Colbert et al. 2022).

The chapter will first discuss key ideas associated with urban planning and residential real estate development practices as they relate to the region's housing recovery. This is contextualised with an outline of the pre-earthquake housing path in the urban region, before examining patterns of housing recovery since 2011. These patterns have been shaped by the Christchurch Central Recovery Plan and the Land Use Recovery Plan, both critical planning initiatives in the post-earthquake management of urban form. Housing developments in the central city, the suburbs and on the urban periphery are discussed in turn.

Urban planning and residential real estate development

Planners are important actors in this narrative: particularly because of the way they attempted to manage the Christchurch urban region's spatial form before and after the Canterbury earthquake sequence. Prior to the earthquakes, planners had put much energy into the Greater Christchurch Urban Development Strategy to rein in the region's sprawling form by permitting

DOI: 10.4324/9780429275562-7

much higher levels of suburban and central city residential intensification than in the past. But the earthquakes confronted them with the urgent political imperative to re-regulate and hasten the residential land subdivision consent process on the urban margins to effect recovery. New land was required on which to build houses to replace those lost in the disaster. Some elements of the pre-earthquake strategy, derived from wide-ranging community consultation and designed to be implemented over many years, had therefore to be relaxed and others fast-tracked. The earlier residential intensification objective had to be partially circumscribed in favour of greenfield development. An additional complicating factor was the implementation of pre-existing commitments to construct a major motorway system to the west, south and north of the city during the recovery period. This cut travel times to the urban periphery, stimulated the growth of satellite towns and steered the urban region back to a well-established path of urban sprawl that is seemingly very hard to step beyond.

By contrast, in the central city and suburbs, the planners did their best to stick with the residential intensification objectives of the pre-earthquake strategy. Suburban intensification was permitted in places deemed appropriate. In the central city, land was set aside and developers were incentivised to build medium-density housing. It was hoped that this would greatly increase the number of people living in the central city. Here, planners were promoting forms of dwelling and ways of urban living familiar in overseas cities, but less so to many Christchurch residents. While in the suburbs and the central city, residential intensification is now more common, 12 years after the earthquake sequence began the central city population has not increased and progress towards building new medium-density housing has been fitful. Planning literature is replete with narratives about complexity, implementation difficulties and unforeseen circumstances and outcomes, particularly when novel outcomes are sought (Berke et al. 2006; Hall 1982; Pressman and Wildavsky 1984). In such situations, planning results are often only ever partial and at times contradictory.

Residential real estate development practices have also been influential in post-earthquake housing recovery. Rather than passively responding to market, regulatory and other institutional arrangements, real estate developers and their financiers actively create the conditions for housing provision and urban spatial form. As Murphy (2019) points out, developers and their financiers operate a set of standard calculative practices with respect to acceptable profit margins that produce the prices that will be paid for land and establish the required house prices needed for development to be profitable. Developers must be able to show their financiers that they are 'capable of achieving benchmark profit rates to secure finance' (Murphy 2020, 1510).

It follows therefore that risk-related developer-financier interactions influence the types and locations of housing that can and will be built. If relatively low risk and profitable, new forms of housing will be constructed. In Christchurch, these include central city and suburban medium-density apartments and attached townhouses, which are particularly attractive to rental investors. Many developers, however, fall back on tried and true, lower-risk design and construction approaches, such as those associated with greenfield and suburban, often owner-occupied, detached houses. The working out of these developer-financier interactions means that urban planning aspirations and housing affordability concerns are not often a key priority for developers and this has become clear in post-earthquake housing development.

Christchurch's pre-earthquake housing path

Detached houses, typically built on relatively large plots of land, have dominated the Christchurch's urban region's development trajectory, material fabric and cultural mores since the city's nineteenth-century establishment (Greater Christchurch Partnership 2018; Memon 2003). The corollary of this situation is that before the earthquake sequence there were relatively few multi-unit attached medium-density houses and apartment buildings in the central city and surrounding urban areas (McDonagh 2017; Vallance et al. 2005). In 2010, there were thus limited opportunities for central city living in Christchurch. At that time, the central city accommodated only 8000 of the city's 376,300 and the wider region's 464,900 residents (2.1 per cent and 1.7 per cent, respectively) (Environment Canterbury 2021a; McDonald 2018a). These central city dwellers lived mainly in low-rent flats east of the business district or in apartments, terraces, townhouses and semi-detached condominia near cultural facilities like the Arts Centre, within easy access of the Ōtākaro Avon River and the 164-hectare Hagley Park. The exception to this medium-density housing story was the area of gentrified moderate-density detached houses in the north of the central city.

Public- and private-sector interactions had an important role to play in these housing patterns and urban management arrangements. Prior to 1989, when New Zealand's local government was significantly restructured, the Christchurch urban region was administered by 11 territorial authorities. Each jurisdiction, led by a separate mayor and council, was responsible for planning its own district, zoning land for mainly detached housing developments and associated service centres, including in some cases suburban shopping malls, all largely initiated and built by private commercial investors. In 1989, these 11 local authorities were amalgamated, the urban region from that time being administered by four councils: Christchurch City, Selwyn and Waimakariri Districts, and Environment Canterbury. The latter is the regional council, and is responsible for not just Christchurch, but the whole of the extensive Canterbury region in the areas of natural environmental planning and regional public transport. It also has a regional urban planning mandate.

As a result of the earlier administrative and land development arrangements, these new councils inherited a dispersed, automobile-dependent, polycentric urban region. This became the defining reality in their planning, housing policy and allied infrastructure and transport provision. The 1990s and early 2000s saw further low-density urban dispersal. By 2004, these local authorities and Transit New Zealand (now Waka Kotahi New Zealand Transport Agency), a central government body responsible for transport planning and provision, had recognised that to create a less dispersed development path, and manage the Christchurch urban region's growth, a long-term and overarching vision was needed. As outlined in Chapter 4, these entities thus initiated the Greater Christchurch Urban Development Strategy with a focus on Christchurch city (minus rural Banks Peninsula) and the eastern areas of Selwyn and Waimakariri Districts (Figure 5.1).

The strategy, formally adopted in June 2007, provided for regional growth and development out to 2041. A proposed settlement pattern was established where 60 per cent of all future growth would be accommodated in intensified development within the existing urban area, with the remaining 40 per cent in greenfield sites (Ministry for the Environment 2008). New arterial and peripheral motorways were planned. Where possible, the construction of medium-density and affordable housing was to be encouraged, particularly in Christchurch city (Christchurch City Council 2010b). At the time of the earthquake sequence the details of the Urban Development Strategy's implementation were well in train, but they, like the urban region itself, were to be greatly disrupted (Greater Christchurch Urban Development Strategy Implementation Committee 2011).

Housing damage, insurance and recovery planning

The earthquakes destroyed 7,860 houses in Christchurch city and Selwyn and Waimakariri districts, and a further 9,100 were made temporarily uninhabitable, requiring major repair or rebuilding (Ministry of Business, Innovation and Employment 2013). The Reserve Bank reported that 170,000 houses (approximately three-quarters of the region's housing stock, and a greater proportion in Christchurch city) were damaged to some degree (Parker and Steenkamp 2012; Wood et al. 2016) and around a fifth had damage exceeding $100,000 in value. As indicated in Chapter 3, the majority of this loss and damage occurred in the eastern areas of Christchurch city and Waimakariri district which were most prone to ground deformation and liquefaction. Christchurch's higher-status estuarine and hill suburbs and those bordering the Ōtākaro Avon River and smaller streams immediately west of the central city, such as Fendalton, were also noticeably affected.

It was also quickly realised that a significant part of the region's rental, affordable owner-occupied, social and retirement housing stock was destroyed or temporarily uninhabitable. This raised concerns about housing security and affordability, particularly for low-income residents (McDonagh 2014). Between mid-July 2011 and early 2013, those who needed to rent, including those who had to find a place to live while their houses were repaired, or who had moved to the city to help with building assessment, demolition and reconstruction, were faced with significant rent hikes, measuring 31 per cent on average across the urban region. In the damaged east, rents rose to a lesser degree, but still averaged 21 per cent (Ministry of Business, Innovation and

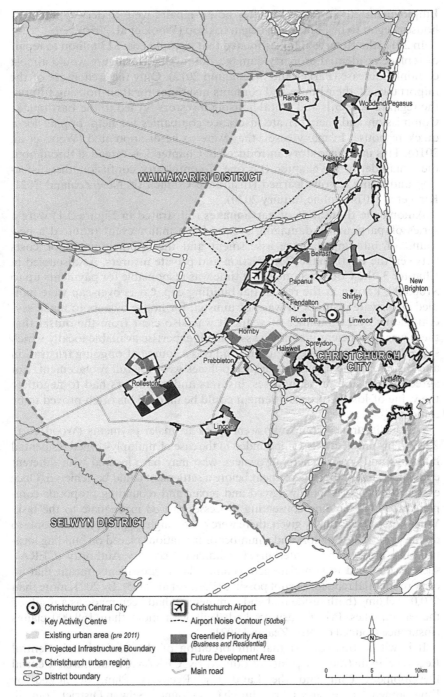

Figure 5.1 The Christchurch urban region.

Source: Greater Christchurch Partnership (2018, 33).

Employment 2013). By 2015, after house repairs were underway and new houses began to be built, rents began to drop (Wood et al. 2016).

In 2012, the Reserve Bank estimated that it would cost $13 billion to repair or replace residential property damage, noting that this figure would almost certainly increase (Parker and Steenkamp 2012). Only too well aware of the importance of their house and contents insurance for their housing futures, the region's residents and rental housing investors turned to the Earthquake Commission and their private insurance companies for help, hoping for a quick response. In many cases, they were to be disappointed (Wood et al. 2016). This insurance story, introduced in Chapter 3, is threaded throughout the remainder of the chapter. It is one of complexity, confusion, dissatisfaction and many lessons learned (Insurance Council of New Zealand 2021; King et al. 2014; Public Inquiry 2020).

Among the thousands of earthquakes (illustrated in Figure 2.1) were a series of particularly damaging events. Each major event required a new round of building damage assessment and the apportionment of costs between the Earthquake Commission and private insurers. As discussed in Chapter 3, the Commission at that time was responsible for payments up to $100,000, known as the residential building cap. Costs over-cap were handled by private insurers. It was not unusual for house owners to make several claims over a number of years. It was also clear from the outset that there was insufficient insurance assessment expertise available locally which slowed the process at the beginning and was a source of ongoing frustration and disagreement. House insurance policies were for full replacement, not agreed value, and in many cases, insurers and customers had to negotiate the value of loss before a settlement could be made. This often proved to be difficult.

Not all insurance companies were quick to make payments (Wood et al. 2016). This problem was compounded in the case of multiply-owned residential buildings and spaces where all owners, who may have insured with different companies, had to be in agreement before a settlement could be achieved. Once each settlement was finally agreed and repair and rebuilding proposals completed, council building consenting processes proved inadequate to the task. When consent was finally given, there were not enough qualified tradespeople to do the work. These factors and a number of limitations placed on building location and progress by Canterbury Earthquake Recovery Authority (CERA), such as the need to complete geo-technical land assessments, meant that an early residential rebuild was not possible (Potter et al. 2015). By 2021, more than 650,000 claims (both residential and commercial) had been made as a result of the earthquakes. Private insurers had dealt with more than 168,000 claims (Insurance Council of New Zealand 2021).

It is within this context that interpretation of the Christchurch Central Recovery Plan (the blueprint), produced by CERA's Christchurch Central Development Unit and the Land Use Recovery Plan, developed by Environment Canterbury, Christchurch City Council, Selwyn District Council, Waimakariri District Council and Te Rūnanga o Ngāi Tahu, working with

Waka Kotahi and CERA (CERA 2013), is essential for understanding housing provision and changes to urban form during the earthquake recovery period. Recovery planning began by incorporating elements of the Greater Christchurch Urban Development Strategy in these new plans for the central city and wider urban region. The blueprint (Figure 4.3) directed the Christchurch City Council to amend its district plan living zone provisions. A condensed and clearly framed central city was planned, as described in Chapter 4. The frame was designed to redefine the central city and provide an attractive and welcoming location for a range of commercial, recreational, cultural as well as residential activities. While the district plan had provided opportunities for medium density residential development throughout the central city, the recovery plan identified, particularly, the east frame for housing, but some also within the north frame (Central City Recovery Plan 2012). It was assumed that the existing areas of downtown medium-density housing would continue to be important. Immediate questions focused on how the new large-scale medium-density developments could be designed and built in a coordinated, attractive and affordable way, and how private sector investment could be harnessed to deliver them.

The future for the urban region beyond Christchurch's central city was laid out in the Land Use Recovery Plan. It covered suburban Christchurch and towns stretching from Lincoln, Prebbleton and Rolleston in the south to Kaiapoi, Rangiora and Woodend/Pegasus in the north (Figure 5.1). This plan was produced recognising that the pre-existing Greater Christchurch Urban Development Strategy with its very strong emphasis on urban intensification and slow land release on the urban margins was not well suited to dealing with the need to build a lot of new housing quickly. Changes to the Regional Policy Statement and respective council district plans were required. The land assigned for slow release in the Urban Development Strategy was to be made available quickly to provide for an anticipated 40,000 new households in both greenfield and intensification areas. It was expected that the private commercial sector would design and build these new housing developments (CERA 2013).

Related transport and infrastructure spending to reflect these land use planning changes was also prioritised in the Land Use Recovery Plan. Consideration was given to revitalising town and suburban centres by identifying key activity centres including satellite town centres in Rolleston and Lincoln in Selwyn District, Rangiora, Kaiapoi and Woodend/Pegasus in Waimakariri District and an array of centres in suburban Christchurch City. Brownfield sites, to be used for residential, mixed-use or business development, were identified, and the plan aimed to permit the building of 8,000–10,000 higher density dwellings in existing, mainly suburban areas, but surprisingly, also planned not to change their character. Overall, the Land Use Recovery Plan stressed urgency and streamlined processes because of recovery needs (CERA 2013). By next considering in turn each of the broad areas of the Christchurch urban region – the central city, the suburbs and satellite towns – it can be seen what kinds of outcomes were achieved.

Central city housing development

Overall, progress in the central city has been both uneven and fitful. In 2012, the city council and the Crown initiated an international design competition for an anchor residential demonstration project, Whakaaturanga Kāinga, to be built on the 8000 square metre Madras Square, a Crown-owned site opposite Latimer Square (Figure 4.3). The objective was to enrol the private sector to create an inspirational exemplar central city medium-density neighbourhood based on sustainable design principles (McDonald 2021a). A New Zealand-Italian consortium won the competition with their 'Breathe Village' plan. The jury praised the proposal for 73 timber-clad three-story house and medium-rise apartments as being 'well-designed and structurally innovative, as well as an affordable and sustainable option'. (Jewell 2014). Sales prices were not indicated but early estimates ranged from $300,000 to $900,000 per unit, at a time when Christchurch's median house price had reached $400,000 (Jewell 2014; McDonald 2013b), thus raising questions about affordability. In November 2015, it was announced that the project had failed due to high-risk market conditions and an inability to secure finance.

In a second attempt to build on the site, the non-profit New Zealand-Australian Ōtautahi Urban Guild was chosen by Ōtākaro Ltd (see Chapter 4) as the preferred developer in late 2018 (McDonald 2018b). The Guild proposed a 150-home affordable co-housing development with a focus on sustainability and communal spaces. It required the city council to carry some of the financial risk, which it did initially in the form of a $450,000 loan. The council ultimately baulked at the Guild's later and much larger $2.8 million loan request, which effectively ended its housing aspirations on the Madras Street site (McDonald 2020a). Ōtākaro Ltd then sold the site to private sector property developers in March 2021. The new owner, a consortium of New Zealand's largest independently-owned house builder and an active Christchurch commercial property developer, proposed to build around 111 homes, ranging from two-bedroom apartments over retail spaces, through to three-bedroom townhouses around a communal garden. Work started in the first half of 2022.

The city council encouraged central city housing development in other ways. In 2018, it initiated Project 8011. Named for the postcode of the central city, it aspired initially to have 20,000 residents living there by 2024, up from 6,000 at the time (McDonald 2018a). To reach this target, it proposed supporting and accelerating housing construction that would meet diverse needs and create highly liveable neighbourhoods. The council thus developed policies to improve the feasibility of new development by providing financial incentives and advice to central city developers (Christchurch City Council 2018). The New Zealand Property Council judged the Project's population target unrealistic, however, noting that high land and building costs, less expensive housing options in the suburbs and a weak apartment culture were impediments to central city residence (McDonald 2018a). Charles Montgomery, a visiting award-winning Canadian author and urbanist, expressed similar views, noting that while the blueprint 'had the right ideas', suburban growth on the urban margins, permitted by the Land Use Recovery Plan, was 'sucking the life from … downtown development efforts' (Doudney 2015, 15).

While these varying hopes, expectations and criticisms were being aired, something interesting, if not entirely straightforward, was happening with central city housing. In September 2014, the Minister of Housing in the National-led government, Dr Nick Smith, had signed a housing accord with the Christchurch City Council aimed at increasing the immediate and longer-term supply and affordability of homes in Christchurch. Consistent with that accord, in May 2015, the Minister announced that Fletcher Living, a national construction company of long standing, had been contracted to build approximately 200 apartments and terraced house on what had been city council-owned land at the edge of the central city but outside the frame. Twenty per cent, or 38 of the new homes were to cost $450,000 or less (Wood and Meier 2015). Further, in July 2015, the Prime Minister announced that Fletcher Living would develop another 1000 townhouses over eight or nine years in central Christchurch, in the east and north frames in an $800 million project (Figure 4.3). The homes were to be built in 14 complexes within three neighbourhoods. As a risk-reduction incentive for Fletcher Living, the land on which each complex was to be built was to be purchased from the Crown progressively as each was constructed (Key 2015). While not confirmed, it was thought that townhouses and apartments would sell for between $400,000 and $900,000 (Wood 2015). It was expected that approximately 2000 people would live in the new housing.

The Fletcher housing development, named 'One Central', and plans for a large neighbouring public park, to be known as Rauora Park, attracted a great deal of interest, not least because they signalled a new beginning for what had become a desolate urban space. But this sense of relief was accompanied by concerns about who might be able to afford the new housing and whether employment growth and the rebuild in the central city would help create sufficient demand (Wood and Meier 2015). The question of investor purchase, where apartments could be listed on Airbnb and other short-stay accommodation websites, was also raised (Law 2020a). Matters were complicated for Fletcher Living by design and commercial issues and it was some years before its One Central housing came on stream (McDonald 2019a). The first of the larger, more expensive units (priced from $1.15 to $1.6m) was available in early 2019 and by June 2019, 'close to six months on the market… [they were] struggling to sell' (Young 2019). Smaller, less expensive apartments sold more quickly, many to investors as rental properties (McDonald 2019b). Construction progress was slow, and by March 2021 only 20 per cent of the planned housing had been built, but with more planned by Fletcher Living. The houses being constructed in late 2021 ranged in price from $675,000 to $1 million (McDonald 2021b). This was at a time when Christchurch's median house price was $700,000.

In a bid to increase the speed of development, in October 2021, Ōtākaro Ltd announced that it was 'excited to be bringing two large central city sites to market for medium/high density housing' (Ōtākaro Limited 2021), which meant that from 2022 Fletcher Living would be joined by other housing developers in the east frame. Beyond the frame, but still in the central city, other property developers have been at work. The Mike Greer Group, for example, has built terraces and apartments, priced between $699,000 and $950,000. The Williams Corporation, a new company formed after the earthquakes to build moderately priced housing, has constructed an array of medium-density housing units in the central city in the price range $300,000 to $975,000 (Williams Corporation 2021). Sales were reported to be 'strong and steady' (McDonald 2020c). This company's business model is not without controversy, however. Some of its apartments were built without motor vehicle garaging in order to keep costs and prices down. Neighbours complained that residents would park their cars on already congested streets (Truebridge 2018), a not uncommon problem in medium-density housing neighbourhoods elsewhere in New Zealand. Others were unhappy with the design quality of some of the new central city housing, a view supported in a report from the city council's urban design team (Law 2020b).

It is clear that by 2022 some of the housing objectives of the central city blueprint were being met, despite early failures, with houses being built and sold in moderately good numbers. On the other side of the ledger, a good deal of the new housing was expensive and not affordable to people on low and moderate incomes, including families with children, and was thus not meeting the Plan's objective of housing a diverse population. The city council's 8011 Project aspiration to have 20,000 people living in the downtown by 2024 was revised to 2028 (McDonald 2018a), but commentators have suggested that this is also not realistic (Zaki 2021). They noted that the population had reached just over 8000, the same as that just prior to the earthquake sequence (McDonald 2022).

Suburban housing

In the years that followed the earthquakes, all suburban houses were assessed for damage by the Earthquake Commission and insurance companies. Many in the west and north of the city, outside areas of liquefaction risk, suffered little, receiving only small payments from the Commission for repairs. In these cases, owners themselves were expected to arrange with contractors to do the work. Damage to paths and driveways was not covered by the Commission, and when these needed to be repaired or replaced, insurance companies contracted appropriate firms directly to do the work. In areas zoned as TC2 and TC3, as described in Chapter 3, being many suburbs closer to the central city, on the estuarine coast and the hills to the east and south of the central city, a lot more houses were significantly damaged. In these cases, insurers took responsibility for repairs or replacement. When houses were replaced, before rebuilding, geotechnical assessments of the ground at individual sites had to be completed, often requiring more sophisticated house foundations than used previously. The day-to-day effect of these assessments and repairs was that the established suburbs were, for some time, alive with workers and their vehicles, and this, combined with above- and below-ground public infrastructure repairs, caused considerable house and neighbourhood disruption.

While the majority of house owners were generally happy with their insurance payments and repairs, it became apparent that all was not well with aspects of the repair process. Lawyers warned that some earthquake-damaged houses were being bought in an 'as-is condition', hurriedly repaired cosmetically, and on-sold concealing major structural damage, leaving the new owners to deal with the mess, and affecting the housing market for years to come (Grimshaw & Co 2014). By 2015, media reports were appearing detailing examples of 'shoddy' insurance-funded repairs, often to floors and sub-floor piles and foundations, sometimes requiring more than one attempt to remediate them (Cropper 2017). By 2017, documents released under the Official Information Act revealed that the Earthquake Commission had received approximately 11,500 shoddy repair claims (Bayer 2017). As these claims were settled, new damage was often found in houses and underground stormwater and sewer drains, leading to further claims being opened up to ten years after the September 2010 earthquake (EQC 2021). This was often both disruptive and stressful for owners. Claims were filed with the High Court seeking remedies. In 2019, the government agreed to spend as much as $300 million to help 1000 Canterbury homeowners caught in a 'legal quagmire' over shoddy earthquake repairs in on-sold over-cap houses (*Otago Daily Times* 2019).

Another group of house owners found themselves in ongoing and seemingly irresolvable disputes with their insurers for significant sums (Hayward 2018). One owner, speaking anonymously, characterised the modus operandi of his insurance company as being one of 'deny, debate, delay', and in this kind of situation, worn down by the negotiation process, some house owners settled reluctantly on terms that they found unsatisfactory. The Insurance Council of New Zealand's position on this situation was that insurance companies were operating in a complex environment where multiple expert reports, commissioned by the Earthquake Commission, insurers and house owners, conflicted with each other, and disputes and delays were therefore inevitable (Insurance Council of New Zealand 2021). In a bid to expedite and resolve continuing long-standing claims, often already before the courts, in 2019 the Labour-led government established the Canterbury Earthquakes Insurance Tribunal, which continued to operate three years later.

While attaining satisfactory insurance results remained the main objective for many people, others, having lost their homes and received insurance or other government compensation payments, or having arrived in Christchurch from elsewhere, were looking for suitable housing. While the repair process unfolded, there was initially some house building in the suburbs, but this tapered off in 2016. It ramped up again to meet increasing demand at a time of low interest rates, and by 2020 a house-building boom was well underway. New house numbers increased by 13 per cent on the previous year, compared with a 5.9 per cent rise nationally. The mainly detached-house Christchurch suburb of Halswell in the south of the city experienced the greatest population growth. Between 2013 and 2018, Halswell's population increased by approximately 7000 people, or 34.3 per cent, putting pressure on local schools. By comparison, Christchurch City's overall population increased by 8.9 per cent in the same period (Christchurch City Council 2021a). Shopping and service precincts were also built with accompanying free car parking as a key feature of this suburban growth.

While detached-house suburbs grew, however, a good deal of medium-density housing appeared in the suburban landscape. Of the 2900 houses consented by the city council during 2020, 'more than half were apartments, flats, townhouses or retirement units' (McDonald 2020c, 1). Many of these were located along arterial routes, near suburban service centres and overlooking parks and other amenity areas, a phenomenon not often seen before the earthquakes (Figure 5.2). Some of this medium-density housing was priced expensively, in the order of $2 million a unit, but most was built and marketed at lower levels, consistent with that found in the central city. And as there, a significant proportion of the new housing had been purchased by rental housing investors. Some buyers were local, but out-of-town investment was also important, with units priced under $550,000 being most favoured (McDonald 2019b). Concerns were raised about the financing of these housing investments, often bought off developers' plans, without knowing how exactly finance would be arranged. One industry commentator characterised the situation as 'over-enthusiastic investment by unsophisticated people, a lot of them out of Auckland' (Edmunds 2017). This reflects the fact that in New Zealand, rental housing investment is seen by many families as a fall-back savings option.

As the medium-density housing sector developed in the suburbs, neighbours of the new housing, and their Members of Parliament, lobbied the city council to tighten the rules regulating this. As in the past, concerns focused on erosion of privacy from being overlooked by neighbours, loss of sunlight and neighbourhood amenity, and increase in crime (Harris and Hayward 2019; Vallance et al. 2005). Pushing back on this view, developers argued that they offered part of the solution to housing affordability and should be allowed to proceed unhindered. The result of high levels of post-earthquake housing supply was that by 2021 housing affordability in Christchurch, both in terms of rents and house prices, while still challenging for people on low and fixed incomes, was better than in other New Zealand regions where housing affordability has become an issue of considerable social and political concern (Figure 5.3). As per the Land Use Recovery Plan, medium-density housing construction contributed to this, but so too did conventional suburban development, where detached houses were getting smaller and being built on smaller parcels of land, leading to another form of residential intensification. Statistics New Zealand data show that an average-priced house approved in Christchurch in 2020 was 127 square metres in size compared with 185 square metres a decade before. This downsizing was partly offsetting the rise in construction costs (McDonald 2021c).

Figure 5.2 Medium-density housing in the suburbs at (top) Wigram and (bottom) in the central city.

Credit: Stuff Limited

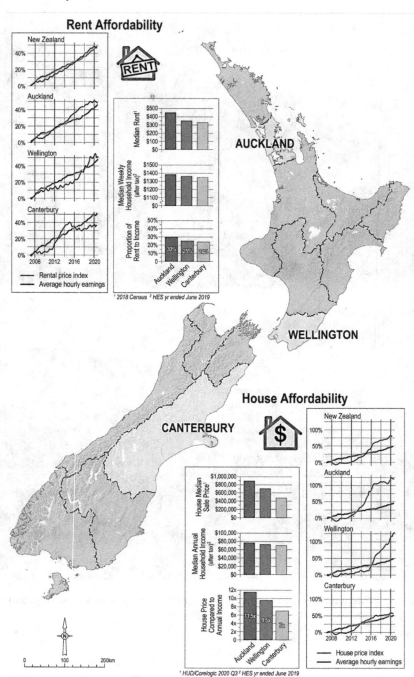

Figure 5.3 Housing affordability in Canterbury and New Zealand.

Source: stats.govt.nz/infographics/the-state-of-housing-in-aotearoa-new-zealand

One exception to this trend in house scale and price was in the high-status suburb of Fendalton, but there too a level of residential intensification became evident. Fendalton is located on the inner edge of the suburbs to the north and west of the central city. While much of the media attention after the initial earthquakes was on the widespread destruction and disruption in the eastern suburbs and around Kaiapoi, local media outlets were silent about damage in Fendalton. But it was clear to anybody watching closely that the houses of the wealthy had also taken a major hit. One commentator suggested that the reason not too many people beyond the suburb fully understood the situation was because 'much of the damage [was] ... hidden down long driveways, behind tall fences and screened by the mature trees and lush gardens that are as much a trademark of the suburb as its whopping real estate prices' (Cropp 2011).

In a locality where properties with stream boundaries were in high demand, and a good number of older houses were built from double brick and other forms of unreinforced masonry, intense shaking, liquefaction and ground deformation had an inevitably devastating influence. The first casualties were the large double brick mansions built in the late nineteenth and early twentieth centuries. They cracked and partially collapsed, and were demolished quickly. Many other houses were still liveable but required major repairs. This situation continued for some years as owners negotiated for often large insurance payments. As in other suburbs, Fendalton saw its share of initial busy house repairs, but then a major rebuild began of big new architecturally designed detached houses, some discussed in *Mansion Global*, an international luxury real estate listing website (Singer 2020). A number of large, very valuable lots, once the site of a single house, were subdivided and infilled by building several detached houses, or groups of townhouses, contributing to further residential intensification in the suburbs. Comparatively high prices were sought for old and new houses in Fendalton, with the median price reaching $1,263,600 in April 2021 (Barclay 2021), and in its 'elite streets' over $2 million (Hawkes 2018).

Two other forms of residential intensification in Christchurch's suburbs are worth mentioning. The first is private sector for-profit retirement villages. These became increasingly conspicuous in the suburban landscape during the recovery period, although this trend was developing earlier with a group of public and private New Zealand and multinational companies offering aged-care facilities and services. In 2021, 100 such facilities provided various forms of medium-density and apartment living in the urban region. Additionally, central government's public housing agency, Kāinga Ora – Homes and Communities, which is one of Christchurch's largest house builders, stresses medium-density living in its design guidelines. During the recovery period, in the suburb of Aranui, for example, this agency built several multi-unit housing complexes (Walton 2021a). Similarly, the Ōtautahi Community Housing Trust, which contracts to Christchurch City Council to manage its social housing stock, also builds medium-density housing. In 2021, the trust completed its 90-unit, Hoiho Lane complex, to replace the Brougham Village which was demolished after the 2011 earthquakes (RNZ 2021).

In these various ways, the objectives of the Land Use Recovery Plan to encourage house repair, and rapid construction on newly released land on the edge of the suburbs, have been achieved. The plan's suburban residential intensification objective, too, has been realised to a noticeable extent. Questions of housing affordability remained for those on moderate to low incomes. It seems likely that cases of residential intensification that are perceived to be poorly designed and located will attract continuing opposition from neighbourhood residents and their political representatives, and that the city council's multi-unit urban design criteria will have to be more strictly applied. Detached houses remain the dominant housing type in the suburbs, but with many new houses being built on smaller lots than in the past.

New housing in the satellite greenfields

One of the most visually spectacular outcomes of the housing recovery has been the construction of extensive detached-housing developments in the satellite townships of Waimakariri and Selwyn Districts. This is reflected in the south and westward redistribution of the region's population since the earthquakes (Statistics NZ 2018). New houses in these developments are being sought by city residents and in-migrants, often with children, who prefer living in single-family housing. These townships comprise a particular kind of landscape of consumption in which houses and a range of retailing and hospitality services are utilised. The houses are significant sites of individual and family leisure, and key to that is the space they offer to use and store an array of leisure-orientated consumer items (Figure 5.4). Commuter travel to and from these satellite towns is also part of their attraction. Heins' (2015) study of the subjective experience of daily travel from the greenfields showed that many residents place a significant positive value on their work journey and related family trips. This is notwithstanding the obvious impact this commuting has on oil use and carbon emissions.

Rolleston and Lincoln townships in Selwyn, the second fastest growing district in New Zealand, stand out as excellent examples of such satellite townships (Figure 5.1). Rolleston is 25 kilometres and less than a 30-minute drive by car from the Christchurch central city at the end of the new southern motorway. At the 2018 New Zealand Census, it had a population of 17,532, up from 8079 in 2013, a 117 per cent increase. The Selwyn District Rolleston Structure Plan indicates that Rolleston's population could reach 50,000 residents by 2075. In 2012, Rolleston had one primary school; by 2021, this had increased to seven. A large new secondary school opened in 2016, with a second on the way. In 2016, journalist John McCrone argued that 'it was time to take Rolleston seriously', pointing out that its public and private service facilities were growing in size and number. These included parks, a new library and community centre, extensions to its swimming pool and many new business premises providing for manufacturing, warehousing, rail and road transport and rural servicing (McCrone 2016a). The construction of a

Figure 5.4 Detached housing in Lincoln: (top) Lincoln Dale subdivision, illustrating
the significance of detached houses for the off-street storage of recreational and other consumption goods. Credit: Alamy; and (bottom) houses
bordering the Lincoln wetland. Credit: Stuff Limited

$85 million shopping mall was announced in 2021. Rolleston is also home to
Selwyn District Council's administration offices. The township is therefore
much more than a dormitory suburb because it also provides a range of
employment opportunities across a range of sectors. The December 2021
median house price at $802,000 in Rolleston was $100,000 more expensive
than that of Christchurch city, but housing remained attractive to buyers
(OPES partners 2022).

Lincoln, on the other hand, is 23 kilometres and 25 minutes by car from the Christchurch central city, and travel times to the city have also been positively influenced by the proximity of the new southern motorway. By 2018, Lincoln's footprint was ten times greater than it was in 1998. Smaller than Rolleston, the Lincoln urban area had a population of 6510 at the 2018 New Zealand Census, an increase of 3690 people (131 per cent) since the 2006 Census. Its structure plan is expecting the population to increase to 10,136 in 2041, but this is likely an underestimate. The town is serviced by two primary schools and one secondary school, and is home to a small specialist university and several Crown Research Institutes. These, along with a growing array of public and private service facilities in its central business district are significant sites of employment. In December 2021, Lincoln's median house price was $812,150 (OPES partners 2022), making the town more expensive to live in than both Christchurch city and Rolleston. Detached house construction continues at pace. As shown in Figure 5.4., some of these houses, built as part of the Ngāi Tahu Property's Te Whāriki development, are located beside a spring-fed wetland planted and managed for stormwater reticulation, and this is a notable part of Lincoln's residential amenity. In 2021, a local developer, Carter Group, applied to Selwyn District Council to have 186 hectares of farmland rezoned, on which it proposed building 2000 houses and developing a small commercial precinct. If consented, this new housing could accommodate a further 5400 residents (McDonald 2021d).

Conclusion

The loss of, and widespread damage to, many houses as a result of the earthquakes was highly disruptive and meant that owner-occupiers and rental housing investors became involved in often protracted engagements with the insurance sector, local government building regulators, and the building industry as they replaced or repaired their properties. Renters were often confronted with limited housing choice and considerable variation in rental costs. Housing availability in the central city, suburbs and urban periphery, and the broader management of urban form, was influenced by the implementation of central and local government recovery plans and the actions of residential property developers. As a result, the suburbs, particularly the outer suburbs, and satellite greenfields displayed a good deal of path dependency when considered in terms of the Christchurch urban region's historical sprawling growth trajectory.

In a process of planning re-regulation, streamlined procedures in the Land Use Recovery Plan, and rapid land subdivision to cater for very significant housing demand, overrode the aspiration of the pre-earthquakes Urban Development Strategy planners for a residentially denser urban region. The resultant release of peripheral greenfield land, that had previously been zoned for later and much slower release, saw developers building what they knew best: a lot of financially low-risk, detached housing. This met the recovery plan objective to build many houses quickly, at relatively affordable prices, which also accorded with the housing preferences of many owner-occupiers. Recent plans for the urban region developed consistent with the National Policy Statement on Urban Development suggest that these detached housing development trends on the urban margin will continue (Greater Christchurch Partnership 2018).

Growth on the margins of the Christchurch urban region was also enabled by new transport links. Originally included in the Urban Development Strategy, these were built primarily to facilitate the movement of commercial transport, and funded as part of the government's Roads of National Significance motorway construction programme (McCrone 2014a; Waka Kotahi New Zealand Transport Agency 2021a). The building of the Christchurch Southern, Western and Northern corridors cut commuting times to and from Christchurch city. The road-building programme, at a cost of $900 million, was announced in January 2010, before the earthquakes, and progressed alongside earthquake recovery. A cycleway programme was also initiated (Waka Kotahi New Zealand Transport Agency 2021b). These cycleways and the construction of new motorways reinforced the long-established trend to sprawl in Christchurch's urban form, but conflicted with the desire for a strong central city embodied in the Central City Recovery Plan (Pawson 2022).

In the central city, and the suburbs built before the earthquakes, a stronger connection between the Urban Development Strategy and Central City Recovery Plan was established. In the suburbs, there is now more residential intensification than previously, and rental investors and owner-occupiers have taken advantage of the new housing. In the central city, developers were encouraged to build medium-density housing, and a number of firms are doing so. But despite this break from past housing trends, progress has been fitful and plans only partially successful, which is consistent with accounts of plan implementation in complex situations where novel solutions are sought. Housing affordability in the central city remains a significant issue. Details are sketchy as to how much of the new housing has been purchased by short- and long-term rental investors. Aspirations to see a much larger, more diverse and vibrant community of long-stay owner-occupiers living in the central city than that in place prior to the earthquakes have not materialised. But it is clear from media reports that many of those who have chosen to rent or buy in the central city are happy with their choice and the urban lifestyle it affords.

6 City centre recovery and commercial property investment

Introduction

Christchurch's commercial property sector was affected dramatically by the earthquake sequence and the building demolition that followed. The loss of workspace dislocated businesses, government and volunteer and community agencies, forcing them to reorganise activities and operate beyond the boundaries of the central city. It also reinforced previous strongly articulated public and private sector commitments to the downtown and refocused pre-earthquake attempts to stem its decline in the face of growing suburban competition, particularly from retailing, hospitality and allied services. There was never any question about rebuilding the central city as a multifunctional urban space in situ, regardless of the difficulties faced. Economic, political, cultural and popular interests coalesced around this approach. Thus local government, the Crown and local property owners engaged in a protracted central city recovery effort which continues to unfold. The objective of this chapter is to outline that recovery process, showing how cooperative, contested and sometimes antagonistic interaction between the Christchurch City Council, Crown agencies, property owners, their insurance companies and the region's residents have led to a partial regeneration in a situation of 'staggering complexity' (Resilient New Zealand 2015).

At the outset, the city council used its resources and expertise to help Christchurch residents envision a future design for the city centre. The council engaged in a public participation project to define that vision and lay out the steps it could take to implement its design. Controversially, as Chapter 4 has outlined, the Canterbury Earthquake Recovery Authority (CERA) then stepped in with its own Christchurch Central Recovery Plan, the blueprint, designed to enrol as many rebuild participants as possible, including overseas investors. To encourage that investment, the Crown promised to commit generously to a range of civic anchor facilities and precincts. Using its statutory powers, and operating under urgency, CERA also limited some lines of property development action and facilitated and quickened others. The Crown's approach was tempered by the knowledge that in a market-centred society such as New Zealand's, commercial property owners and their building insurers, over which they held limited sway, were going to be key actors in any form of central city recovery.

DOI: 10.4324/9780429275562-8

Property investment decision-making and calculative practices

After the earthquakes, city administrators and the news media were alive to the potential of capital flight from Christchurch if property owners decided not to reinvest in central city recovery (McCrone 2011). They were influenced by a popular view that property investment decision-making and behaviour is economically rational: that private property owners seek to maximise profit in all circumstances and act accordingly (Diaz 1999). While it is true that post-earthquake Christchurch did not seem a propitious place to invest, and there was some capital flight, in the event, most local commercial property owners remained and re-invested in the city. This boundedly rational behaviour contained important strategic and affective elements (de Bruin and Flint-Hartle 2003).

Its strategic elements reflected calculative practices that are locally generated and quite different from those performed by national and trans-national institutional commercial property investment companies (Halbert et al. 2014). Such companies are often owned by multiple shareholders, sometimes listed on national and international stock exchanges, and hold large high-value property investment portfolios comprising clusters of individual properties. Their investment decisions are influenced strongly by a fiduciary duty to minimise risk and maximise returns for shareholders. The people who manage these companies are guided by algorithms that use investment data collected by specialist agencies, and decisions are made significantly in economically rational 'spread-sheet' financial terms (David and Halbert 2014). Such data are not readily available for small provincial cities like Christchurch, and so most institutional property investors do not have the information that would enable them to decide to invest. Additionally, in such places, there are usually very few high-value commercial properties, in the order of $NZ200 million or more, that would be attractive to institutional investors. The few that do exist in Christchurch are suburban shopping malls and these are held in a combination of local and institutional ownership (McDonagh 2021). A good example is Westfield Mall in the suburb of Riccarton. It is owned by Scentre Group, Australia's 15th largest listed company by market capitalisation, valued in 2019 at approximately $A20.68b (Gibson 2019).

Local investors, on the other hand, who prior to and after the earthquake sequence, represent the majority of commercial property owners in the central city, hold land and buildings that, when compared with those owned by institutional investors are of relatively lesser value. These local investors are strongly connected to regional economic activity and allied social networks, and manage their investment portfolios in direct support of particular end uses such as office services, retailing and hospitality. They do not make decisions in purely financial terms. Their decisions about what gets purchased, built and maintained are in part influenced by relationships with financiers, but because many properties are held closely, often multi-generationally, with limited debt, financier-property investor relationships are not as important as they might otherwise be. These investors do not have a fiduciary duty to shareholders and are able to 'take a punt' with their investments in at least slightly riskier ways than those demanded of institutional investors (McDonagh 2021).

When seen in terms of affect, or emotional connection, limited capital flight and commitment to the rebuild in Christchurch reflected local investors' strong senses of place. This is a key finding of research by Ikenna Chukwudumogu who in 2017 interviewed 20 'informed property stakeholders' about their perspectives on commercial property and earthquake recovery in central city Christchurch. They included two public sector managers, three property consultants, 14 property owners and one property finance manager. On the basis of this fieldwork, it was apparent that local commercial property investors, who of necessity became developers after the earthquakes, identified strongly with the city and were emotionally connected to its future (Chukwudumogu 2018a, b; Chukwudumogu et al. 2019). While clearly focused on a reasonable return on capital, at least at some stage in the future, they displayed many of the characteristics of boundedly rational property entrepreneurs reported by Levy et al. (2021). These are strongly place-attached, risk-orientated local business people who imagine positive futures for their hometowns. They are committed to neoliberal market-centred and individualistic values and place-based property development that advance their own and others' business and community interests.

CERA, the city council and later Ōtākaro Limited, the Crown's property developer, thus had to implement their plans alongside and in conjunction with these often well-resourced property owners who had their own agendas to pursue. In general terms, each on their own – local government, the Crown and property owners – believed they had the tools needed to achieve their objectives; but in a chaotic and complex post-disaster environment these tools showed themselves to be imperfect and not completely controllable (Perrow 1986), and in their struggle to gain some semblance of control, the central city rebuild slowly and unevenly worked itself out. In order to contextualise this, the next section considers the pre-earthquake situation of commercial property owners in the central city, and describes their responses to earthquake loss.

Responding to loss in the central city

Prior to the earthquakes, a wide range of private and public services were located in the Christchurch central business district, making it a significant place of employment, hospitality, recreation and education. The concentration of long-established cultural facilities such as the Canterbury Museum, the Christchurch Arts Centre and the Anglican Cathedral set in, and alongside, public squares and other open spaces, made the central city important as a destination for international tourists. The people who worked and played there did so in buildings owned mainly by local, but also some national and international entities, among which were represented a significant group of local wealthy individuals, families and trusts (McDonald 2013a). Some of the local people were owner-occupiers, others investors, the latter making an income from letting commercial property. These locals had dominated the commercial property market since the 1980s and 1990s, when many institutional property investors left the central city for other more profitable locations in New Zealand and overseas, for example, the much larger cities of Auckland and Sydney, because of the relatively low returns to be obtained in Christchurch (McDonagh 2021).

Low investment returns were an element of the widespread recognition that before the earthquake sequence parts of the commercial central city were economically depressed, despite ongoing local government and private sector efforts to rejuvenate it by attempting to attract more people and investment. As one indicator of this, Chang et al. (2014, 8) suggested that 'up to forty per cent of the available floor space in the central city may have been unoccupied prior to the earthquakes'. While having considerable architectural and heritage character, many low-rise buildings constructed in the late nineteenth and early twentieth centuries from unreinforced masonry were nearing the end of their useable commercial lives (McDonagh 2017), although this was a matter for debate between owners, the city council and heritage advocates. From the 1950s, multi-storey steel reinforced concrete office towers had been constructed, and these larger buildings were tenanted to varying degrees by office services, hospitality, accommodation and retailing businesses. The tallest was the 23-floor Pacific Tower, completed in 2010.

Retailing had become a notable challenge. The size of the retail area of the central city was far larger than required, particularly in the face of sustained competition from suburban shopping malls offering free car parking. As was the case for office space, shopping space in the central city was under-utilised: many shops existing there because rents were low, or the occupiers were also owners. Local journalist John McCrone (2016b) suggested that at that time 'Christchurch felt simultaneously over-shopped and yet underwhelming', noting that 'There was plenty of activity, but of a cheap and unfocused kind.' McCrone quoted a retail analyst who had in the past been hired by the city council in an advisory role; in his view 'Christchurch had quantity but not quality. Much of the retail had low productivity and a lot of churn'.

Despite this, in his 2017 interviews, Chukwudumogu (2018a, b) and Chukwudumogu et al. (2019) heard accounts of the feelings of loss, shock and grief experienced by building owners after the 2010 and 2011 earthquakes. All 12 buildings owned by one informant had been damaged or destroyed, including a hotel and two restaurants which he ran himself with 40 staff. He had found the situation traumatising, and was surprised to lose all his holdings at once. By the time of the interview, he had only one remaining staff member. Similar responses to the loss were reported by those who rented and worked in these damaged and destroyed buildings, and the requirement that they vacate their premises caused many challenges. The Crown responded by offering subsidies to assist employers to pay employees at a time when businesses were unable to operate, as described in Chapter 4. These were in addition to the business interruption insurance payments received by some companies (Poontirakul et al. 2016). After the September 2010 earthquake, 64 per cent of businesses had to close at least for a short time. More than one per cent shut down permanently. But after the February 2011 earthquake, the situation was more severe, with 11 per cent closing permanently and the others closing their doors for a median period of 16 days (Potter et al. 2015).

The central city was cordoned for safety reasons by the Civil Defence authorities after the September 2010 earthquake but this lasted only a short time. But immediately after 22 February 2011, Civil Defence reinstated the cordon on a larger scale, encompassing an area of 387 hectares and named the 'CBD red zone' (Chang et al. 2014). With checkpoints manned by police and army personnel, and taking various diminishing shapes, the cordon was to last until 20 June 2013, by which time it covered just a few city blocks and had become the responsibility of CERA who had renamed the cordoned area the 'CBD Rebuild Zone' – although popularly it was still called the 'red zone' (Figure 6.1). The effect of the cordon was that initially, building and business owners were denied access to their premises while structural and safety assessments were conducted and building demolition took place. While in hindsight it is widely agreed that the cordon was necessary and effective, at the time it caused a good deal of stress to those whose freedom of access it limited.

The cordon operated to considerable effect. Six thousand businesses were forced to leave the area within it, affecting 50,000 central city jobs. Not surprisingly, this situation resulted in considerable volatility in the local commercial property market. Businesses and central government departments migrated to the edge of the city. They often took up tenancies in vacant industrial and warehouse properties after owners rapidly converted them to offices, retail and hospitality spaces to take advantage of the high demand and tenants' willingness to sign rental contracts for long enough periods to ensure tenure (Chukwudumogu et al. 2019). These were premises often not of the quality and size preferred (Moricz et al. 2012). The new tenants were to be working in these places for some time, and in late 2021, some were still in place. In another development, a suburban-based property entrepreneur and brewer, Alasdair Cassels, took advantage of the flight to the suburbs and built a multi-unit facility designed to be rented to small retailing and hospitality businesses that had lost their premises in the earthquakes. His property development, The Tannery, is on a former industrial site in the suburb of Woolston, and houses 62 small retailing, hospitality, craft and entertainment businesses, some once in the central city and others having different origins (Stylianou 2013).

As in the case of housing (Chapter 5), insurance was a key concern for building owners after the earthquakes. But unlike housing, the Earthquake Commission had no role in commercial property insurance and so insurance companies and their overseas reinsurers were the centre of attention. Reflecting the fact that insurance coverage for the Canterbury earthquakes was notably higher than in similar jurisdictions overseas (Wood et al. 2016),

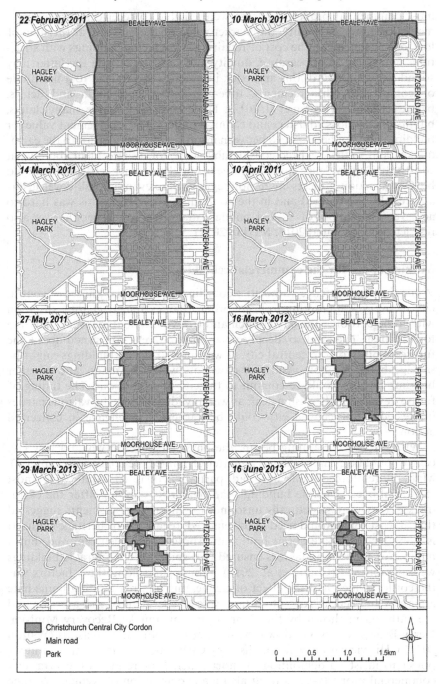

Figure 6.1 Reduction of the Central City cordon, 22 February 2011 to 16 June 2013.

most commercial property owners were eligible to receive insurance payments for either indemnity, the market value of the building at the time of
loss, or reinstatement, the cost of repairing or replacing buildings with structures equivalent in appearance, quality and function. While the latter may
sound straightforward, this approach in practice was to lead to a multitude
of complexities, disagreements, time-consuming negotiation and delays,
before settlement (Brown et al. 2013). There were, for example, many claims,
but insufficient loss adjusters and claims handlers. The replacement value for
buildings had not been agreed between insurer and assured prior to the earthquakes, and so there existed considerable scope for disagreement about what
constituted 'equivalence' and how large or small insurance settlements should
be. Construction projects could not begin until relevant all-risk-cover insurance could be obtained and in the continuing aftershocks this was hard to
achieve. The supply of skilled builders and other tradespeople was also
restricted (Grollimund 2014; Merkin 2012).

Making space for the commercial rebuild

While the Civil Defence authorities coordinated the initial phases of central
city recovery – attempting to bring order to an extremely chaotic situation
within the cordon – the city council and Crown initiated recovery planning
for the area. They did so, deeply uncertain and very concerned about the
future viability of the central city. It was unclear which buildings might be
salvageable, quite how insurance payments would be made, and whether
building owners would re-commit to the city once paid out. The timing of
future developments was unclear, as was any idea of how the various parties
interested in the central city could be harnessed to engage in concerted
action.

When recovery planning began, the physical space on which a new city
centre could be established was a matter of immediate concern (Chapter 4).
Many unreinforced buildings had partially collapsed, and while most
steel-reinforced concrete buildings had remained standing after the earthquakes, they were structurally unsound. Engineering assessments suggested
that at least some would need to be removed, regardless of heritage status. It
is a matter of debate about how many should have been removed, but a
number of interacting factors ensured that some likely repairable buildings
were demolished (Chang et al. 2014). Prominent among these factors was
the cordon which had the effect of delineating an area needing immediate
attention and action. Also important were a general air of risk aversion and
the enthusiasm shown by the Canterbury Earthquake Recovery Minister,
Gerry Brownlee, 'to pull down the old dungers' for safety reasons (Walton
2021b). His view was operationalised by CERA because it had the statutory
power to insist on demolition in short order. Low returns on central city
commercial property investment also made demolition convenient to some
property owners, particularly when linked to their insurance arrangements
(Marquis et al. 2017).

The demolition process picked up quickly, growing from 21 demolitions in March 2011 to a peak of 124 in July that year. That was approximately 30 a week or four a day (Gates 2012a, 2015a). The demolition zones within the cordon were busy, dusty and dirty places. Fascinated locals and visitors congregated along the cordon fences, enthralled as they watched and photographed large cranes swing wrecking balls, and high-reach excavators equipped with tools such as pulverisers, shears, breakers and grab attachments, slowly deconstruct tall buildings. Figure 6.2 illustrates the March 2011 demolition of the art deco St Elmo Courts, a Category II listed heritage building constructed in 1930 (with the earthquake-strengthened Christchurch City Council office building standing undamaged behind). These demolished buildings were turned into easily transportable rubble, comprising many materials. Trucks carrying debris away from the central city were a common sight on Christchurch's arterial roads for many years. 850,000 tonnes of material were taken to the Burwood Resource Recovery Park on the edge of the city for recycling, or dumping in two 25-metre-high hills (Law 2019a). An estimated one million tonnes of clean fill were carried to the Port of Lyttleton to allow the Port Company to reclaim 10 hectares of the harbour for its new container storage space, a policy that proved environmentally controversial (Wood 2011).

Figure 6.2 Demolition of St Elmo Courts in March 2011.

Source: http://ketechristchurch.peoplesnetworknz.info/canterbury_earthquakes_2010_2011/
images/show/18773-st-elmo-courts-288-montreal-street-22-march-2011-2. Licensed under a
Creative Commons Attribution-NonCommercial-ShareAlike 3.0 New Zealand (CC BY-NC-SA
3.0 NZ)

The number of demolitions was not to dip below 20 buildings a month, or five a week, until June 2012, but by February 2015 had declined to two a month. By that time, 1240 buildings within Christchurch's four avenues had been demolished, with approximately the first 1000 being razed by the end of 2012. These comprised 70 per cent of the central city business office, retailing and hospitality stock. Perhaps reflecting the confusion surrounding the measurement of this phenomenon, the Insurance Council of New Zealand (2021) has published slightly different demolition figures, saying that 1354 commercial buildings needed to be demolished in Christchurch – 826 in the central city and 528 in the suburbs. Whatever the exact number, among the buildings demolished were 147 listed with Heritage New Zealand Pouhere Taonga, although the city council count is over 200 as it classifies heritage buildings differently (Walton 2021b). This, as Chapter 4 indicated, was a very contentious issue.

The large scale of the demolition had a dramatic effect on the central city landscape. Figure 6.3 illustrates this in an aerial view of a portion of the central city commercial area – the full city block bounded in the east by Madras Street. The top photograph taken on 24 February 2011, two days after the major damaging earthquake on 22 February 2011, shows some collapsed buildings, notably the CTV building in the bottom right of the block. The bottom photograph was taken five years later on 20 February 2016 of the same area after demolitions and before significant rebuilding in the east frame. The Christchurch Transitional (Cardboard) Cathedral is evident in top right of the image and the 185 Chairs memorial to the people who died in the earthquake sequence is in the lower right.

As central city building demolition proceeded, the city council produced its draft Central City Recovery Plan, as described in Chapter 4, submitting it to the Minister for Earthquake Recovery in December 2011 (Christchurch City Council 2011). The council had consulted on its plan and this attracted 15,500 comments and 2,800 submissions. While many submitters were happy with the proposed direction for the central city, a number were scornful. Among the latter were local commercial property owners and their representatives. They complained that they had not been adequately included in the creation of plans that would very directly affect them and require their property, money and skills to implement (McCrone 2011). Angus McFarlane, owner of more than 10,000 square metres of the city's commercial floorspace, and a member of a family that had held commercial property in Christchurch for over 60 years, told the submissions hearing that 'the city was in despair before the earthquakes, with a failing central economy and diminishing values'. He claimed councillors had made a mess of the city before the quakes, and the draft plan 'proves you have got it wrong again'. McFarlane described the 'obstructive, dictatorial and impractical' plan as a 'rubbish pipedream' that terrified potential investors (Sachdeva and Carville 2011).

He was partially supported in this view by Connal Townsend, the chief executive of the property owners' advocacy organisation, the Property Council of New Zealand, who said that while property owners and developers

Figure 6.3 A portion of the central city seen from the air in (top) March 2011 and (bottom) March 2016.

Sources: 2011: https://data.linz.govt.nz/layer/51932-christchurch-post-earthquake-01m-urban-aerial-photos-24-february-2011/ and 2016: https://data.linz.govt.nz/layer/53454-christchurch-0075m-urban-aerial-photos-index-tiles-2015-2016/

supported the city council's plan, they were 'terrified' by the details of the proposed regulations, and what they interpreted as extinguishing property owners' existing-use rights (Sachdeva and Carville 2011). While no specific mechanism for doing so was outlined in the plan, it was clear that land title amalgamation and some form of compulsory land acquisition was a likely outcome, described later by journalist Rob O'Neill (2012) as the 'elephant in the room no-one is talking about'. Townsend's view was that such an approach, combined with the general uncertainty about and volatility in the local property market, would greatly increase the risk that central city property owners would reinvest outside Christchurch (McDonald 2012). While resisted by Mayor Bob Parker, these views about the council's plan and the potential for capital flight were held strongly among some property owners and likely influenced CERA's evaluation of the council's plan and its response to it.

Controversially, Minister Brownlee rejected the city council's plan overall, and the authority set about producing its own Christchurch Central Recovery Plan, the blueprint (Figure 4.3). A notable feature of the blueprint's design-led approach (Brand et al. 2020) explicitly confronted the difficult matter of amalgamating existing property titles in order to speed up rebuilding, and particularly to provide the large blocks of land required for the anchor projects and other big scale developments. The likely use of CERA's power to compulsorily acquire land in this amalgamation process had begun to be openly discussed by senior figures in the commercial property sector after the submission of the council's plan but prior to the development of the blueprint (O'Neill 2012). It was clear to them, as it was to Warwick Isaacs, the Christchurch Central Development Unit's newly appointed Director of Earthquake Recovery, that some rationalisation of the central city's variably sized lots held in 3000 land titles was required.

Isaacs wrote in the tender document seeking a consultant to prepare the blueprint that such a situation

> means that it is difficult to undertake development at a variety of scales sought by the market, and in particular, larger-scale developments that could gain from efficiencies and support multiple tenancies. There is little evidence of landowners consolidating their land holdings to create larger scale projects.
>
> (O'Neill 2012)

How to incentivise that consolidation was a key question in a context described by Hamish Doig, managing director of property brokerage Colliers Christchurch, where many properties were held by a significant number of investors whose debts had been extinguished on receipt of insurance payments (O'Neill 2012). The newly minted blueprint also focused on attracting property investors and developers from beyond Christchurch to the central city rebuild, particularly from overseas (McCrone 2014b). Consistent with this aspiration, the Development Unit established a new agency called 'Invest Christchurch' which was charged with developing an international investment marketing campaign (Wood 2012).

As the blueprint was of great interest to local investors and property owners, Christchurch law and real estate firms were quick to offer their advisory and technical services to property-owning clients. A good example is {mds} *law*, a Christchurch legal firm, who produced a recovery plan newsletter (2012), opening with the statement: 'The redevelopment of the central city of Christchurch is an opportunity unlikely to be seen again in several lifetimes. While the redevelopment project promises many exciting investment opportunities it also creates much uncertainty for affected land owners.' In order to inform their clients of a likely future for central city properties, the newsletter writers first described the blueprint's proposals as

> creating a smaller Central City Business Zone made up of a "Core" of commercial, retail, hospitality and residential developments girded by a "Frame" of open space and beyond that a large format retail area. This central Zone was to be 40 hectares in extent, down from 92 hectares prior to the earthquakes.

They then described the proposal to build 17 publicly funded anchor projects valued at \$4.8 billon: these to be a range of civic facilities and other initiatives designed to kick start the central city rebuild and attract people and businesses to the centre (Figure 4.4).

Property owners were informed in the newsletter that the Canterbury Earthquake Recovery Act enabled the Minister to fast-track the process of designating land for these anchor projects and approximately 840 properties were to be purchased to enable the construction of the four highest priority projects: the convention centre, Ōtakaro Avon River precinct, the east frame and the metro sports facility. This land was to be acquired either voluntarily with the price negotiated with the owner, or compulsorily, where compensation would be determined under the Act, and possibly at less than market value. Further properties would be required for the development of the remaining anchor projects. This acquisitions process unfolded during 2012 and early 2013. For property owners who wished to engage in voluntary purchase this heralded a complex and time-consuming set of interactions with the Christchurch Central Development Unit in the calculation of property values. Factors involved varied from property owner to property owner and included possibly: insurance, mortgage, securities and tax issues; the status of commercial tenancies; the actual location of the property *vis-à-vis* the blueprint; and land swap arrangements. Where multiple properties were held in different parts of the central city, complexities and uncertainties were compounded.

Apart from this approach to land acquisition, the blueprint supported a rapid approvals process for all applications for new buildings. But allied to that, a number of restrictions and limitations were to be placed on redevelopment in the Central City Business Zone. New buildings had to cover 100 per cent of the site on which they were located, have a maximum height of seven or eight stories and a minimum height of two stories, with car parking up to a maximum of 50 per cent of the premises. In the retail precinct, it was planned to consolidate multiple small sites into minimum lot sizes of 7500 m^2 (one-third of an existing block). The vehicle for achieving this was the fast-track Outline Development Plan and owners of small land titles were encouraged to collaborate with neighbouring landowners, or face compulsory title amalgamation and/or acquisition, so that a plan could proceed expeditiously. Again, this placed property owners in often complex and difficult situations including trying to work out their best way forward: perhaps to sell to a neighbouring property owner or engage in a development partnership, and if so, create the most appropriate legal structure for advancing their chosen option.

Property owners and investors respond

In the face of the planning tactics outlined above, commercial property owners were left to mull over their options. Some chose to invest their capital elsewhere, but it is hard to know the scale of this phenomenon (McCrone 2012a). A good example of a well-established local property owner who took this option is Angus McFarlane whose response to the city council's draft recovery plan was discussed above. In 2013, it was reported that he was living in Houston, Texas, having taken millions of dollars out of Christchurch to invest in oil and gas exploration in the United States. In an interview, he said that he 'didn't do it lightly, but that it was a culmination of everything'. The 'everything' in this case involved continuing dissatisfaction with the work being done by the council and CERA and the fact that the anchor project and precinct designations 'had cost him half the value of his properties' (McDonald 2013c).

The Christchurch Central Development Unit's hoped-for influx of foreign property investment in Christchurch's rebuild failed to materialise (Conway 2014), and so the insurance payments owing to property owners – which in fact were sourced from overseas reinsurers – proved to be the resource that underpinned strong local property owner involvement in the rebuild (McDonald 2013d). The size of the payments to any particular individual reflected the number of properties held and the details and type of their insurance cover, with the amounts in aggregate running to $10.86 billion by March 2021 (Insurance Council of New Zealand 2021). Owners received settlements ranging from hundreds of thousands to many millions of dollars. These insurance payments, supplemented by borrowed and private funds, financed the Christchurch private commercial rebuild. Notwithstanding the widespread concern about the loss of heritage buildings, this process allowed old and sometimes dilapidated structures to be replaced with new and functionally better ones.

On the basis of their research, Chukwudumogu (2018a, b) and Chukwudumogu et al. (2019) argue that local property owners were strongly place-attached and passionate about Christchurch and its future. They had an opportunity to re-invest in a place they knew well and their sense of local connection contributed to their decision, despite few opportunities for immediate profit. As well-resourced and intimately connected locals, these property owners were well positioned to dominate the rebuild. Thus, cautiously optimistic, they proceeded to pursue property investment plans by evaluating emergent opportunities in a situation where there was less land to work with than previously. They were to be faced with moments of worry that this decision was the wrong one. Some of them indicated that re-investing may not have been the best decision when judged purely in financial terms; it was their desire to succeed and contribute to exciting new ventures in the reconstruction of their home town that overrode initial doubts (Chukwudumogu et al. 2019).

For several years after the earthquakes, these property owners expended much effort on the time-consuming complexities of building damage assessment, engaging with myriad allied stakeholders and creating plans for new buildings. Adroitness and flexibility were key requirements. The uncertainty that accompanied this busy round of activity was in part influenced by the tardiness of the blueprint's anchor project roll-out, and it took until 2015 for some clarity to be achieved about that process. The fact that many of the building owners' former business tenants had relocated to the suburbs and were on long leases, raised questions about the short- to medium-term tenanting prospects for any new building they might construct. The data to apply to systematic forecasting of returns, yield and financial viability using standardised methods were not available. Investment decision-making was instead influenced by intuition, past experiences and interaction with others in the commercial property investment community. In this situation, property owners found themselves taking on new and unexpected roles. Where originally, they had been property investors, making a living from letting buildings built by others, they now had to become speculative property developers. They did not, therefore, always understand the processes and challenges of undertaking commercial property developments and this exposed them to risks not before encountered (Chukwudumogu 2018a and b; Chukwudumogu et al. 2019).

This lack of property development and project management expertise resulted in some being unable to manage their projects effectively. Some reported that they had incurred unnecessary costs because in their inexperience they had permitted the construction of over-engineered buildings (Chukwudumogu 2018a and b; Chukwudumogu et al. 2019). In this, and similar situations, their building projects took longer to complete than initially anticipated, and when completed sat untenanted. A good example is the story of the St Elmo Courts rebuild, the pre-earthquake version of which is illustrated in Figure 6.2, and an early candidate for demolition. In February 2015 it was reported that its owner, Richard Owen, was 'bitter and angry about the whole city and deeper in debt than ever' (Gates 2015b). He had rebuilt a six-storey building to very high seismic standards. Having only indemnity insurance he had received a payment of approximately $2 million. The original tender for the building had been $10.5 million but the final cost was $17 million. At completion the building was only half tenanted which was 'pretty nerve wracking and very stressful' (Gates 2015b).

Owen's experience was compounded by the existence and effect of the central city cordon. Many of the investor-developers discussed above held property within the cordon and this was one of the factors that slowed their progress. But property owner-developers such as Owen, who operated outside the cordon, were able to advance their building plans early. The result was a relatively quick edge-of-city-centre commercial rebuild focused mainly on Victoria Street, Durham Street and Cambridge Terrace to the north and west

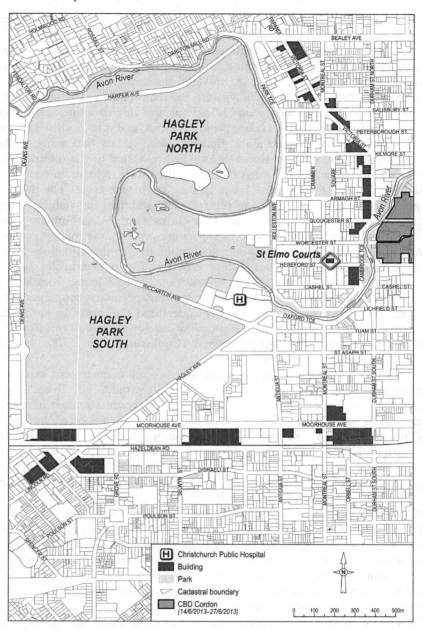

Figure 6.4 Location of edge-of-centre post-earthquake commercial buildings.

of the city centre, and on Moorhouse Avenue and in the area surrounding lower Lincoln Road, in Addington, south and south west of the city centre (McCrone 2014b) (Figure 6.4). The combined result in 2015, as Owen found to his cost, was an over-provision of office space that was difficult to tenant.

The progress of the central city rebuild

The beginning of the central city rebuild was slow, affected by landowner disputes about land title amalgamation which proved to be much harder to achieve than initially thought (McCrone 2012a). While deemed 'sluggish' in April 2015 by journalist and rebuild commentator Liz McDonald (2015a), the private commercial rebuild gained momentum as the cordon shrank (McDonald 2015b). To facilitate office construction the Crown commissioned private developers to construct four office buildings to house approximately 30 of its agencies. This brought a significant group of workers into the central city daily. With respect to shopping, after a halting start and several attempts to create viable Outline Development Plans, discussed above, the billion-dollar-plus City Mall was rebuilt in the retail precinct harnessing the combined effort of several local wealthy business people, some with multi-generational commercial property investment and development interests (McCrone 2015a; Steeman 2012). These included Philip Carter, Nick Hunt, Antony Gough and Tim Glasson.

Created by other younger less well-established property developers, Richard Peebles, Kris Inglis and Mike Percasky, the innovative landscapes of consumption centred on the retailing and hospitality facilities Little High Eatery (McDonald 2017a), which opened in May 2017 and the Riverside Market (Law 2019b), which began operating in September 2019 proved to be very popular. Equally successful, has been the Terrace hub of bars, restaurants and office space on Oxford Terrace beside the Ōtākaro Avon River to the west of the city centre (Kenny 2021a; Mead 2017). These were conceptualised by brothers Antony and Tracy Gough in 2013, and opened progressively, slowed by financing difficulties, through late 2017, with the popular pre-earthquake Viaduct Bar joining them in January 2021. These developments are discussed in more detail in Chapter 9.

In September 2017, Christchurch's revitalised central city was 'still seen as fragile' (McDonald 2017b). In this context, the managing director of a commercial real estate firm reflected on the slow progress of the Crown and city council's major anchor projects (Figure 4.4) and their perceived importance in attracting people to the central city. Intended to be completed during 2017, none of these major projects was delivered on schedule. He said that

> It's really nice to see vibrancy in the spaces created. I just hope there is continued investment by the Crown and the council, and quickly. People in the private sector have made a considerable investment – now it's time for the public sector to deliver on some of its promises.
>
> (McDonald 2017b)

The pace of the commercial rebuild plateaued in 2016–17. Chukwudumogu et al. (2019) point out that in 2016 the office vacancy rate in the city centre was volatile and rents declined into 2017 as a result of office oversupply. Property owners competed to attract tenants from the suburbs by offering incentives such as rent-free periods or fit-out contributions. Smaller operators were still denied access because they could not afford even these lower rents.

A concern expressed early in the recovery process was that the new seismically strengthened central city office buildings and shops, which were more expensive to build than those of the past, and fewer in number, would demand much higher rents than those paid prior to the earthquakes. To a significant extent, this has been true. Corporate and government offices and global brand stores like H&M and Country Road dominate the central city because they can afford to pay. Thus, in many respects, these stores replicate shopping available in suburban malls, but without the free car parking. There is some anecdotal evidence that at least a few landlords levied rents consistent with the funds they had borrowed, rather than the total cost of their new buildings, seeing their insurance payments as a windfall. This was in reality a subsidy on tenants to help them operate in the central city. Despite this, high set-up and operating costs limited margins for central city businesses and in combination with unaffordable rents this, for some, was a cause of their failure. The beneficiaries of this situation were often the next tenants who had lower set-up costs and were therefore able to establish more financially sustainable businesses (McDonagh 2021; McDonald 2020d).

Much of the building construction was done by New Zealand firms, but in order to do so, they required labour sourced globally. In 2015, immigration rules were changed to allow 5000 rebuild workers to enter the country on Essential Skills visas. The arrival of thousands of migrant workers, primarily from the United Kingdom, Ireland and the Philippines injected increasing levels of multiculturalism into the city (Statistics NZ 2018). A number of problems also arose. The warning issued by the Engineering, Printing and Manufacturing Union that this influx of workers would create housing problems was borne out (Jones 2015). Welfare agencies reported that some workers, particularly Filipinos, were exploited by employers and landlords and found themselves living in cramped conditions and paying high rents (Garces-Ozanne et al. 2022; Meier 2014; Stylianou 2015).

Conclusion

The central city commercial property recovery is still unfolding. There is potential for much more to be done. While estimates have varied (Gates 2018), in December 2020 approximately 20 per cent of central city land, 68 hectares in over 450 sites, was empty. Many sites were used as temporary car parks and the cause of public concern, being both visually unattractive and often unconsented (Law 2020c; Walton 2021c). That month, the council approved a plan to 'disincentivise holding vacant land, by using its capacity to levy rates differentially (Law and McDonald 2020). Allied to these issues, a number of central city property owners failed to repair or sell their buildings for redevelopment. Dubbed the 'Dirty 30' by the council, these unoccupied and deteriorating structures were identified as barriers to the central city rebuild in 2017 (Walton 2020). They were either precariously positioned or encroaching on public areas, or partially demolished or derelict. A few were the subject of insurance disputes. Council staff sent the owners warning letters, offering guidance and outlining enforcement possibilities if action was not taken. By 2018 the list had been reduced to eight properties, partly because it had mistakenly included a number of significant heritage buildings (Church 2017). In 2021, city councillor and regeneration working group chair James Gough, a relative of the Gough family of Christchurch commercial property developers and investors, identified the remaining six buildings on the council's list as the 'worst of the worst' (Mitchell 2021).

Greater levels of central city commercial activity were deemed to be the solution to these problems. In late 2021, a report from CBRE New Zealand, a branch of the global commercial real estate and investment firm headquartered in the United States, suggested that such an increase in activity seemed likely. CBRE indicated that the Christchurch central city office vacancy rate had dropped below the pre-earthquake level and that this would stimulate a new round of construction. Consistent with earlier media reports (McDonald 2021e), key actors, CBRE noted, would be a small number of local and offshore land owners either taking on the development role or selling to those who would. Lead times to completion for these developments were likely to be in early 2024 which suggested the likelihood of a squeezed market and increasing rents in the meantime (*The Press* 2021).

Notwithstanding these issues, given the high level of uncertainty and volatility that prevailed after the earthquakes, the progress made in the central city has been significant. This has been due to the adroit and flexible approach taken by property investors turned developers, as they and Crown agencies, the Christchurch City Council and other stakeholders struggled to create innovative opportunities for recovery. Much remains to be done, however, and it will be some time yet before the scars of the demolition process are less evident. Questions remain about the likelihood of attracting 20,000 residents to the central city (Chapter 5). The Greater Christchurch Partnership (2018) expects that over the next 30 years the central city will gain an additional 40,000 jobs, resulting in over 75,000 people working there. The contrast between the situation in early 2022, and this expectation, is stark, and it is not clear quite how such a transformation is to be achieved. This is because, alongside these developments in the central city, the number of attractive, easily accessible and competitive suburban and greenfield retailing and allied service and employment centres continues to increase as Chapter 5 has shown, and therefore the tensions that existed prior to the earthquakes with respect to central city's economic viability are still evident. No easy solutions are in sight.

7 Voluntary and community sector responses

Introduction

Voluntary and community organisations played a significant role in the post-disaster recovery of Christchurch. In the weeks and months following the quakes, a number of well-established organisations, many of which were faith-based, provided temporary accommodation, meals and counselling to affected residents, as well as assisting people to access government support. Several resident-led organisations also arose, energised by concerns regarding the approach of government agencies and private insurers. The Student Volunteer Army emerged as a distinctive form of volunteer-based social movement, with large numbers of students coming together to offer a range of practical assistance.

This chapter examines these three broad forms of voluntarism – that of faith-based organisations, resident-led initiatives and the Student Volunteer Army – and seeks to assess their wider significance. Each form is described in terms of the support it entailed and its intersection with the agendas of state recovery agencies. The position of voluntarism in relation to politically conservative aspirations for rebuilding the city and more progressive forms of place-making is also considered. To support the analysis, the chapter begins with an overview of voluntary welfare provision in Christchurch prior to the earthquakes. Although the terminology used to refer to the voluntary sector varies internationally and includes 'third sector' and the 'not-for-profit sector', the language of 'voluntary welfare sector' and 'voluntary and community organisations' is preferred here. This is in keeping with the vernacular in New Zealand academic and policy circles (Nowland-Foreman 2016).

The background to post-disaster voluntarism in Christchurch

Prior to the earthquakes, the voluntary welfare sector in Christchurch was often associated with the activities of several large and relatively professionalised organisations, many of which had roots in particular denominations of the Christian faith. These organisations – including Anglican Care, Methodist Mission, Presbyterian Support, Catholic Social Services and the Salvation Army – employed significant proportions of paid staff and often

DOI: 10.4324/9780429275562-9

held service provision contracts with government funding agencies, such as the Ministry for Social Development and the Ministry of Health. While they made some use of volunteers – in their neighbourhood charity shops, food-banks and drop-in centres, for example – their core services were typically provided by salaried social workers, psychologists, counsellors, policy advisors, accountants and funding application managers. Alongside these large and broadly focused organisations, a number of small- and medium-sized voluntary welfare organisations were also active in the city. These tended to concentrate on specific issues, such as food insecurity, tenancy advice or counselling, and often relied more heavily upon volunteers.

The neoliberal reforms that took place in New Zealand from the mid-1980s significantly altered the conditions in which these voluntary organisations operated (Tennant et al. 2008). From 1984, economic and social restructuring led by a Labour government saw increases in unemployment and social hardship. Faced with reduced taxation revenue and a burgeoning welfare expenditure bill, a centre-right National government elected in 1990 then implemented stringent cuts to state welfare, reducing the value of benefits and tightening their eligibility criteria (Boston et al. 1999). This retrenchment compromised the material security of some of the country's most vulnerable citizens and saw a considerable expansion in the numbers seeking support from the voluntáry sector.

From the mid-1990s, the National government introduced a funder-provider system to manage its service provision engagements with voluntary welfare organisations. The funders (government agencies) would contract with providers (voluntary organisations) for the delivery of particular services, with various associated performance indicators and reporting requirements (Boston 1995). This new arrangement was problematic for many voluntary organisations as it entailed significant administrative burdens. There was also concern that government funding contracts might compromise their capacity to critique or operate independently of government policy (Nowland-Foreman 2016), as has been observed elsewhere (Baker and McGuirk 2021; Milligan and Conradson 2006; Wolch 1990). In recognition of these problems, a Labour-led government (1999–2008) created an Office for the Community and Voluntary Sector in 2000, with the aim of improving relations with the community and voluntary sector. This Office worked to ameliorate some of the more onerous and bureaucratic aspects of service contracting while endeavouring to foster a more collaborative working relationship between government and the voluntary sector. The arrangement was not to last, however, as a subsequent National-led government disestablished the Office in 2015, effectively reinstating the emphases of the earlier contractually oriented funder-provider model.

By 2010, immediately prior to the earthquake sequence, voluntary organisations were a key provider of welfare and social services in Christchurch, spanning housing, child and family support, mental health, community development, residential aged care, budgeting and financial advice, migrant and refugees services and citizens' advice (Conradson 2008). Many of the established organisations held significant service provision contracts with government agencies. A number of small- and medium-sized organisations also provided welfare and care services, in areas such as counselling and psychosocial support. These services were often funded by community grants, donations and client fees, and were typically more financially vulnerable than the larger entities.

Community and voluntary organisations often take up important roles after disasters, with responses that encompass emergency and medium-term relief, as well as advocacy and longer-term work that seeks to ameliorate distress and disadvantage (Quarantelli and Dynes 1977). This was certainly the case in Christchurch, where the response encompassed organisations well prepared for disaster recovery work (eg. the local branch of the Red Cross) as well as those without any particular expertise in this regard.

Faith-based organisations: established and new forms of voluntarism

Reflecting their existing involvement in welfare provision in Christchurch, faith-based organisations played a significant role in responding to immediate social needs following the earthquakes. Parsons (2014) describes how Christian churches and organisations were closely engaged in distributing essential supplies, establishing response and recovery centres, providing pastoral care as well as caring specifically for children and older people, and delivering broad psychosocial support. Other faith groups also played their part, including the members of the city's two mosques, although their contributions were less publicly visible at the time. This was to change following a terrorist attack on the mosques in March 2019, however, one outcome of which was to significantly raise public awareness of the extensive community involvement of members of the Christchurch Muslim community (Uekusa et al. 2022). Although a number of the larger faith-based organisations suffered earthquake damage to their buildings and the homes of their staff and volunteers, many made immediate use of their institutional resources to deliver food parcels, establish emergency shelters and provide a range of other forms of immediate care.

A number of congregation-centred responses also rose to prominence in the aftermath of the quakes. For example, Grace Vineyard Church had already established a Compassion Trust in 2003 to provide meals and budgeting advice for local people. This kind of under-the-radar contribution was given wide-ranging publicity when the Beach Campus of Grace Vineyard emerged as a major hub for post-quake food distribution, community support and legal advice in the badly affected area of New Brighton (Figure 7.1). The church leader concerned described this welfare response in the following way:

> In any disaster, there is going to be a large group of people who actually have the energy, capacity and hopefully the spiritual eyes to put their own needs aside to serve and to bless others ... you are best just to try to do something ... just hold on for the ride and make something happen.
>
> (Harvey 2014, 223–4)

Figure 7.1 The Beach Campus of Grace Vineyard church.

Credit: Michal Klajban, Wikimedia Commons. Creative Commons Attribution-Share Alike 4.0 International license.

Alongside the buildings and volunteers, and a certain form of ethical citizenship, Harvey's description also highlights the improvisational character of those who were prepared to 'try to do something' and to 'make something happen'. As explored in Chapter 8, this sense of experimental innovation enabled by the rupture of the earthquakes appears to have been a common feature of post-disaster Christchurch. Indeed, the achievements of Grace Vineyard in helping to co-ordinate wider community responses were repeated by other congregations elsewhere in the city. In Richmond, for example, the Delta Community Support Trust – established in 1995 as an initiative of the North Avon Baptist Church – also emerged as a major community hub after the earthquakes, becoming not only a focus for emergency relief but also operating a wide-ranging community development centre serving disadvantaged people in the eastern suburbs (Burn 2020). Similarly, the congregation of South West (formerly Spreydon) Baptist Church built upon its previous experience of community work with at-risk youth, sole parents and those needing mental health support to engage with local residents and their communities in the aftermath of the earthquakes. This community and ecumenical cooperation blossomed through the recovery period, and helped to spawn and consolidate some 40 different charitable trusts in the neighbourhood and further afield (Jamieson 2018; Table 7.1).

It has been suggested in Chapter 1 that alternative possibilities for action that had previously been submerged, covered up or excluded might be afforded new prominence after the social, political and environmental upheaval of a disaster. The rupture of the Canterbury earthquakes does appear to lend weight to this suggestion given the emergence of new kinds of

Table 7.1 Examples of charitable trusts associated with South West Baptist Church, Christchurch

Name of Trust	Nature of service
Building Blocks Community Trust	A fully licensed Early Childhood Education Centre
Cross Over Trust	Works with primary school-aged children and their families to support resilience and well-being
Cobham Street Trust	Community housing provider for individuals and families in need, with a focus on the elderly
Sarona Trust	Residential service for adults recovering from mental illness, with accommodation and support programmes
Kingdom Resources Trust	Budgeting and employment support service
Stepping Stone Trust	Supported accommodation for adults, youth and families affected by mental illness

Source: www.swbc.org.nz/ministry/associated-trusts/

welfare knowledge, priorities and even organisations. In this regard, some welfare needs became more apparent because of the earthquakes. For example, the established faith-based organisations, the congregational community support trusts and a number of counselling organisations recognised the need for additional mental health support services in the city. Burn (2020) estimates that the scale of mental health issues responded to by the Delta Trust rose by 70–80 per cent in the decade following the earthquakes. Apart from the uncertainties and, for some people, the fear generated by the earthquakes, one of the main sources of anxiety was having to deal with insurance issues, as other chapters (3, 5 and 10) make clear. This anxiety was exacerbated by the collapse of expectations that high levels of personal and social insurance would provide security against disaster (Chapter 2).

The earthquakes also re-emphasised the issue of homelessness in the city. Some of the housing stock vacated in the red zone and rendered unusable in the eastern suburbs included affordable properties, social housing and low-cost rental dwellings and their absence created further homelessness and housing poverty, in addition to that already existing in the city. Major service providers such as the Anglican City Mission (Haworth 2019) reported that responding to the earthquakes highlighted the need for additional services – such as a women's night shelter and a men's daytime programme – both of which were subsequently introduced. In addition, there was a broader and more visceral appreciation amongst some Christchurch citizens of the day-to-day conditions endured by those experiencing housing-related distress and homelessness. Over shorter or longer periods of time, the insecurity, uncertainty, displacement and deprivation resulting from the loss of home, however temporary, threw new light onto the issue of homelessness and highlighted the need to provide services for its victims.

This increased prominence of particular social needs in the post-quake city was matched by faith-based communities refashioning their responses to those needs. Interestingly, for many of the established faith-based organisations, this largely meant returning to a position of 'business as usual' (Hawkey 2020). To some extent, this development reflected a desire to maintain a professional and caring service for the city's marginalised people. As Haworth (2019, 159) summarises in the context of the Anglican City Mission:

> What the Mission has to offer the people who come to it is much more than a smart almost-new building. It offers well-trained and motivated people, all committed to a high standard of caring. It offers services that are relevant, well-resourced and carefully honed to provide affirming care.

Although such priorities do not preclude innovation, they are often articulated in the context of co-funded partnership projects with government agencies. For example, the Methodist Mission had become the lead agency for the collaborative implementation of the government-funded Housing First programme in the city, which sought to address the housing needs of people experiencing long-term homelessness. In doing so, it has added significant strands of service to its longer-term commitment to provide housing for older people (Hawkey 2018, 2020). It has also partnered with other agencies in the Mana Ake (Stronger for Tomorrow) initiative, which seeks to ensure that every school in the city benefits from the services of a kaimahi (well-being worker) to help deal with anxiety, depression and other mental health issues amongst school children following the earthquakes.

The core services of these established faith-based organisations were augmented by a number of community-based development initiatives, the locational pattern of which has shifted from experimental projects in a few communities to a more geographically comprehensive coverage. Some of these organisations have not relied on faith-based motivation and resources for their development: an example would be the Aranui Community Trust Incorporated Society (ACTIS 2022). This initiative encompasses a broad range of well-being services – including housing and legal advice, social connectedness interventions, early childhood education, affordable fruit and vegetable provision and community sports, among others – and was prompted by interactions between the local community, city council and Housing New Zealand (now Kāinga Ora). Elsewhere, however, faith-based organisations have contributed significantly to community-based service provision across the city. What perhaps could be thought of as a limited number of relatively independent service silos prior to the earthquakes has grown into an important community presence in the post-quake city.

It has been argued that the earthquakes powerfully reinforced the need for faith-based congregational social action to be seriously collaborative (Harvey 2012). Not only did ecumenical cooperation during the earthquake emergency help to dissolve denominational boundaries, but community cooperation became central to establishing trust between church and neighbourhood. For example, Harvey's (2012) account of the Grace Vineyard response in New Brighton emphasised the fundamental requirement for co-operation between faith-based actors and others, including local politicians and police leaders. In these kinds of ways, the event of the earthquakes helped to fashion a less organisationally centric and more community-oriented approach to faith-based social action, leading to a rise in community-centred services that augmented the core services offered by the established faith-based organisations.

As the recovery of Christchurch continues, the established faith-based organisations have returned to their prominent position as major trusted service providers, along with new services developed in response to some of the forms of need caused or made visible by the earthquakes. In a welfare landscape with relatively few services provided directly by the state, faith-based voluntarism always risks co-option as a pseudo-state agency (Cloke et al. 2013). However, the evidence from Christchurch suggests that there is a determination on the part of faith-based actors to maintain an independent voice and conscience when partnering with government agencies, regardless of any use of state funds. Other faith-based initiatives have sought deeper integration with local communities through community-based development projects. These smaller and often more localised initiatives have the capacity to become key hubs for responding to local and neighbourhood-based needs. For example, Delta Trust's position in Richmond places it in an area where significant gentrification is occurring but also where social housing developments have led to a concentration of socially marginalised people. The situated understanding of local community among staff at Delta also enable a focus on less visible needs, such as those experienced by isolated older people or by people with mental health needs.

Locally integrated activity has brought with it two significant benefits. First, there is a particular expression of compassion-based care, which is not exclusive to faith-based motivation, but which often seems to draw inspiration and resourcefulness from the wellspring of faith communities and their practical orientation to reach out to others in need. Such practices can form part of a broader antidote to neo-liberalised cultures of renewal in the post-quake city. Secondly, as faith-based groups act on this ethical imperative to join with others to 'do something about' the needs they see around them, they bring a capacity for experimentation and creativity that is perhaps more difficult to achieve within longstanding established institutions. This capacity reflects the wider opportunity for an alternative cultural politics that has been afforded by the rupture of the event of the earthquakes (Cloke and Dickinson 2019; Cretney 2016, 2019).

Resident-led advocacy and activism

Alongside the work of established voluntary organisations, a number of resident-led initiatives arose in the aftermath of the earthquakes, often energised by concerns around justice, fairness and equity. These initiatives sought to shape government policies and practices, taking up both collaborator and critic roles. Some had a pan-city orientation while others focused on particular neighbourhoods. As illustrative cases, this section considers three examples. The first, the Māori Recovery Network, was formed in the immediate aftermath of the February 2011 event, and drew on existing tribal networks and support structures. The network was led by Ngāi Tahu, the tribe with mana whenua (customary authority) in Te Waipounamu, the South Island. The second example, the Canterbury Communities' Earthquake Recovery Network (CanCERN), arose after the September 2010 earthquake to represent the interests of residents in disrupted neighbourhoods. The third, the Wider Earthquake Communities Action Network (WeCAN), was established a year later in October 2011. Each of these groups sought to assist residents, in overlapping but also different ways. They were part of a wider set of resident-led organisations, several of which played influential roles in the city's recovery.

In the immediate aftermath of the February 2011 earthquake, Ngāi Tahu actively coordinated a response effort across the city. Much of this work was channelled through the Māori Recovery Network, with participants drawn from the tribe, key government departments, Māori social service providers and Māori wardens (Kenney et al. 2015). The Māori wardens were closely involved in offering practical support, with individuals drawn from a national network of approximately 900 people. Guided by whanau (extended family) networks and manaakitanga (an ethic of care), these wardens made several hundred in-person visits to Māori households in the eastern suburbs of the city, offering material assistance (eg. temporary accommodation arrangements, food parcels and drinking water), advice and information, service referrals and psychosocial support. The close attention given to Māori cultural values, language and worldview in offering this support was particularly significant for local Māori (Lambert 2014).

In his reflections on the Māori Recovery Network, Mark Solomon, the Kaiwhakahaere/Chairperson of Ngāi Tahu at the time, emphasises how tribal connections with local Māori enabled the network to offer support that was well calibrated to people's actual difficulties. He contrasts this with the protocol-oriented approach of Civil Defence, which was generally less flexible and less well attuned to local needs. Although the Māori Recovery Network was actively involved in household visits, particularly in the eastern suburbs, it was not until eight days after the February 2011 earthquake that Civil Defence met directly with representatives of Ngāi Tahu, despite their status as mana whenua. Solomon nevertheless argues that the earthquakes eventually led to improved relationships between Ngāi Tahu and government agencies such as CERA and the Christchurch City Council. In his words,

> [t]he earthquakes were a tragedy but they also led to a strong Treaty relationship for Ngāi Tahu. The earthquakes gave us a unique opportunity to design a cityscape that acknowledges our shared past with our colonisers, our shared experiences and our common future, which certainly wasn't there beforehand.

> (Solomon and Revington 2021, 87)

As a second example, CanCERN arose from the concerns of several resident associations and community groups, with an aim to facilitate '[f]ull community engagement in recovery processes and to work in partnership with recovery agencies' (CanCERN 2014). To that end, it sought to

> share accurate information regarding the earthquake recovery process with communities and affiliated supporters via a regional network; identify and advocate for community-based solutions and future vision; establish engagement partnerships with key-decision makers in the earthquake recovery process; [and] promote communication and engagement processes that are inclusive rather than divisive.
>
> (Miller 2012)

From the beginning, CanCERN aspired to work alongside government agencies and other emergency response organisations in a collaborative and non-antagonistic manner. As its early documents put it, 'our aim is to partner, not criticise from the sidelines' (Miller 2012). In addition, CanCERN did not wish to function or be perceived as a 'lobby' group. The intention instead was to facilitate a two-way flow of accurate information between residents on the one hand and the government and insurers on the other.

Supported by a steady stream of funding from charitable trusts and foundations, CanCERN developed a series of initiatives. One prominent example was the formation in 2012 of a community information hub. Operating out of a shopping mall in the eastern suburbs, this hub brought together representatives from 18 different organisations, including the Residential Advisory Service (offering free legal advice), Ngāi Tahu, the Human Rights Commission and local and central government agencies (eg. the city council and the Earthquake Commission). This arrangement facilitated valuable opportunities for face-to-face conversations, enabling people to make progress on issues that might otherwise have taken multiple emails and telephone conversations to resolve. To keep in contact with the community, a public Facebook page was created for information and regular updates (Figure 7.2).

A further project, the 'Find and Fix' campaign, was announced in April 2014. This sought to organise temporary repairs to more than 900 badly earthquake damaged homes, to remedy their weather tightness before winter. In September 2014, a new initiative called Breakthrough was developed in collaboration with Southern Response (the government agency formed to take over the claims made against Allied Mutual Insurance, a private insurance company that went into liquidation). Staff working for Breakthrough sought to help residents progress their insurance claims. As Dickinson (2018, 627) notes, 'CanCERN staff acted as brokers who could access legal and technical advice, facilitate meetings with insurers, bring managers and technical staff to the table and ensure there were agreed actions for progress'.

These conversations were not always easy, given the levels of frustration and anger among city residents regarding insurance claims. But CanCERN

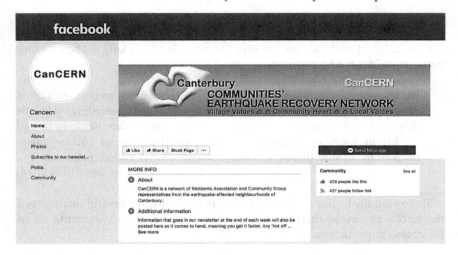

Figure 7.2 The CanCERN FaceBook page.
Source: https://www.facebook.com/Cancern-447406772071865/

described its work as seeking to transcend the antagonistic politics of 'us' (disaffected residents) and 'them' (government agencies, such as CERA, and insurance companies) (for the origins of these disputes, see Chapters 3 and 4). It sought to create 'a space of cooperation and collaboration' between different actors and agencies, so as to assist residents whose insurance settlements or home repairs had stalled or reached an impasse (CanCERN 2014). For CanCERN, the effectiveness of its work depended upon building close relationships with key staff in agencies such as CERA and the Earthquake Commission. Some people were concerned that CanCERN was too close to the government, however, arguing that its collaboration with CERA compromised its ability to advocate effectively for residents and communities. Given the generally slow progress towards resolving insurance claims and the complexity of red zone negotiations, these critics wanted a more distinct and adversarial approach. Vallance (2015) argues that CanCERN staff were in fact consciously treading the line between collaboration and critique so that they could 'stay in the tent', or be in a position to influence government actors from the inside, as a trusted partner, rather than having to do so from the outside. Dickinson (2018) also observed that CanCERN staff regularly assessed what was politically acceptable. They adjusted their approach to the situation at hand, in ways they thought would have the best chances of success.

In late 2015, ahead of the planned disestablishment of CERA in April 2016, the organisers of CanCERN decided to bring their work to a conclusion, with a finishing date in February 2016. As outlined in Chapter 4, the key actors in the city's recovery were shifting, and important decision-making powers were to be transferred from CERA to a new set of organisations (primarily, Regenerate Christchurch and Ōtākaro Limited). In the final CanCERN newsletter, headed 'the Space of Discomfort', the editorial acknowledged the challenges but also the rewards of working at the interface between residents and government.

> We also understand that the very term "collaboration" may be perceived as a negative. We have recently been accused of being quislings (Google that one!) but we are actually proud of the space we are working in and are taking every opportunity to work with the agencies and organisations to make sure there are better outcomes for the stuck residents.
>
> (CanCERN 2014)

Following the notice of CanCERN's intended closure, several members of the public expressed their appreciation of its work, with comments on its Facebook page such as:

> Thanks so much CanCERN, goodness knows where the heck we'd be without your years of hard work. I really appreciate the effort to keep the community in the rebuild ... Thank you!

> CanCERN folk, you guys have been so amazing and certainly helped me keep relatively sane through our fight. To see a group of people come together, to ask questions and to actually care and communicate with the average joe has been a huge blessing ... You asked the hard questions, and in such a skilful way that you were rarely 'shut-out'. That is no easy feat.

These remarks were typical of the positive community response to CanCERN's work. The government agencies, including CERA, apparently also valued the opportunity to work with CanCERN, having recognised the importance of engaging with resident-led organisations early in the recovery process (Dickinson 2013).

In its final months, three core staff from CanCERN announced they were forming a new organisation called Breakthrough Services (taking its name from one of the previous CanCERN initiatives, mentioned above). Breakthrough Services would continue to support homeowners to work through earthquake insurance issues, adopting a collaborative approach to advocating for their needs and rights. This reconfigured, smaller organisation also offered consulting services, drawing on the experience that its staff had gained working for CanCERN. The aim was to help 'agencies who would like to make their communications and processes more people-centred. In the complex disaster recovery environment, engaging meaningfully with people is the key to success' (Breakthrough Facilitation 2019). This kind of service was illustrative of the capacity building that occurred as a result of working in challenging post-quake circumstances; for the city, it was positive that this expertise survived the closure of CanCERN and thus remained available to local residents across Christchurch.

In contrast to CanCERN, WeCAN took a more confrontational and adversarial stance in its efforts to achieve better outcomes for quake-affected residents. Led by Mike Coleman, an Anglican priest and Avonside resident, WeCAN was particularly focused on red-zoned residents and the difficulties experienced by people living in the eastern suburbs. As Coleman explained:

> We aim to publicly highlight the injustices and issues affecting residents following the Canterbury earthquakes. We will openly challenge decisions, policies and practices that disadvantage a community's recovery from the earthquakes, and actively promote and support equitable, just and visionary solutions for all … We are not politically aligned and will challenge the policies and practices of any political party or government department where they disadvantage residents or communities. However make no mistake: we are political with a small p, [and] we intend to take action for change.
>
> (Voxy 2011)

Through a series of public rallies, letters and legal challenges, WeCAN sought to contest what it regarded as the top-down 'command and control' ethos of CERA and the unsatisfactory processes and communications of the Earthquake Commission and insurance companies, such as IAG (Figure 7.3). Many WeCAN participants doubted whether the Minister for Earthquake Recovery, Gerry Brownlee, was actually listening to Christchurch people. As Coleman put it:

> [i]t makes you wonder if the Prime Minister has forgotten us. He has left us with an earthquake recovery minister who says he is unaware there is a crisis … How can he be the recovery minister when he fails to listen to the people most in need?
>
> (Gates 2012b)

Figure 7.3 Reverend Mike Coleman, Leader of WeCAN, at a public gathering in September 2012.

Credit: Stuff Limited.

WeCAN sought a more democratic recovery process in which the needs and perspectives of residents were given greater attention. Its efforts enjoyed widespread support of residents from the eastern suburbs, as well as from several elected city council representatives – including the Mayor, Lianne Dalziel, who had been advocating for a more even distribution of power between the city council and the Crown – and faith-based organisations such as the Methodist Mission (Gates 2012b).

Energised by this ethos, WeCAN actively challenged each of the main recovery agencies. It submitted multiple requests to the Earthquake Commission for information regarding its decisions and procedures, including the process it had followed to appoint the national building company Fletchers as the main repair agency, and also the progress of the repairs themselves. Through these requests – and by making public aspects of the responses it received – WeCAN sought to hold the Commission to public account. In relation to CERA, WeCAN requested an extension to the time-frame for finalising red-zoned land and home deals, and asked that red zone decisions be reviewed by an independent third party. The timeframe for the red zone settlements was eventually extended, after pressure from multiple quarters, including WeCAN.

In terms of legal challenges, WeCAN worked with Quake Outcasts, a group of 68 uninsured red zone homeowners, to develop a submission to the United Nations Human Rights Council. The submission questioned the human rights acceptability of residents effectively being forced – through their financial inability to move elsewhere – to live in damaged and unre-paired housing while their insurance claims languished in an unresolved state. In a separate process in July 2016, WeCAN applied to the High Court of New Zealand regarding the Quake Outcasts. The Crown had offered a settlement of only 50 per cent of their land value and nothing for their homes (Quake Outcasts 2016). The Court found that the Crown's approach had been unlawful, and required it to compensate the Quake Outcasts in a man-ner equivalent to other red-zoned home owners. Although the approach taken here was very different to the collaborative style of CanCERN, it was clearly effective.

Emergent voluntarism: the Student Volunteer Army

The day after the September 2010 earthquake, Sam Johnson, a Law and Politics student at the University of Canterbury, contacted the Christchurch City Council asking how he might contribute to the recovery effort. He was politely told that 'there is no process in place for your offer of help' (Warnock 2014). Unsatisfied with this response, Johnson reached out to a group of 200 friends on Facebook, asking who might be interested in offering practical help to badly affected residents in the eastern suburbs. To his surprise – as his invitation had been relayed through a range of networks – he received around 2500 offers of volunteer help. This was the start of what became known as the Student Volunteer Army (Johnson 2012).

Following the establishment of an informal leadership team and some rapid logistical work, students were coordinated into teams that shovelled silt and liquefaction out of hundreds of homes and gardens in the eastern suburbs, while also offering residents food, drinks and information about relevant support organisations. This work was facilitated by the deft use of social media, a flexible organisational structure, and a strong desire to respond to the practical needs of disaffected residents (Nissen et al. 2021; Tatham 2011). When the February 2011 quake struck, the Student Volunteer Army was able to leverage its experience and learning from September 2010. It scaled its efforts and activities up considerably, recruiting over 11,000 student volunteers who then became involved in a variety of helping initiatives across the city (Figure 7.4).

After the September 2010 earthquake, there was initially some tension between the Student Volunteer Army and the key emergency response organisations of Civil Defence, the Canterbury Earthquake Recovery Authority and the city council (Johnson 2012). The students were regarded as well-meaning but generally unqualified individuals whose efforts might generate additional complexity and create safety risks (Cowlishaw and Mathewson 2012). After Sam Johnson publicly described some of difficulties they were experiencing in these respects, a more harmonious way of working was developed. Part of the solution was a division of labour, whereby the students would focus on the high volume but relatively low skill tasks that were a necessary part of the recovery. After the September 2010 and February 2011 quakes, the Army thus focused on immediate material assistance, including shovelling silt and liquefaction from affected properties. More complex issues, such as dealing

Figure 7.4 Members of the Student Volunteer Army shovelling silt.
Credit: Stuff Limited.

with broken water mains and sewage infrastructure, or tasks with potential health and safety risks, such as working alongside damaged buildings, were left to qualified professionals.

The kinds of difficulties the Student Volunteer Army encountered when attempting to work with existing emergency response organisations in Christchurch have been observed in other post-disaster settings (Stallings and Quarantelli 1985; Strandh and Eklund 2018). Established organisations typically have well-structured response protocols that may, in some circumstances, prove to be quite efficient. However, these approaches are often less able to accommodate novel community initiatives or emergent groups that also wish to participate in the recovery effort. This was demonstrably the case with the Student Volunteer Army: the scale and dynamism of its activities caught many of the existing recovery agencies by surprise. The situation eased when the Student Volunteer Army created a website, took steps to register as an official club at the University of Canterbury and was later incorporated as a registered charity; these elements of formalisation made it easier for more established organisations to collaborate with it. The formation of a governance board also gave external agencies some assurance that suitable oversight was in place for handling financial and other forms of donations (eg. equipment such as tools, wheelbarrows and software, much of which was gifted by businesses from across the city).

Despite the initial uncertainty of recovery agencies regarding how best to collaborate with the Student Volunteer Army, its community efforts received an overwhelmingly positive reception (Ensor 2011). In a paper co-authored with Sam Johnson, Nissen et al. (2021) nevertheless caution against uncritical eulogising of the organisation, noting that its success in part reflected the extensive social networks and cultural capital of both its founder and the original leadership team of six students. This was coupled with the skilful use of social media to leverage a high level of participation from a large group of generally fit and healthy young people. Being able to connect to the infrastructure and branding afforded by the University of Canterbury and the UC Students Association was also helpful (Carlton et al. 2022).

One might assume that the Student Volunteer Army was created on the basis of a clear vision, which was then implemented through a set of corresponding actions. Nissen et al. (2021) emphasise that this was not the case, but that a good deal of experimentation was instead involved. This sometimes entailed 'throwing a lot of different things against the wall, and seeing what would stick', in the words of one interviewee (Carlton et al. 2022, 42). Such an approach is typical of what Gardner (2013) observed among similarly emergent voluntary organisations in disaster relief on the Gulf Coast of the United States. This orientation towards learning through iterative 'doing' mirrors the key elements of what has become known as design thinking (Brown 2009). At the same time, the Student Volunteer Army leadership team demonstrated a significant capacity for productive reflection. This supported the organisation to adjust its approach in an incremental fashion through the sequence of major earthquakes in February, June and December of 2011 (Carlton and Mills 2017).

Having acquired a degree of organisational experience and capacity through the earthquakes, the Student Volunteer Army was well positioned to respond to other disasters. In April 2011, its founding members Sam Johnson and Jason Pemberton travelled to Japan to contribute to recovery efforts following the 2011 Tohoku earthquake and tsunami (Stuff 2011). Within New Zealand, new groups of student volunteers were then directed to assist people in the aftermath of the 2016 magnitude 7.8 Kaikōura earthquake, the 2019 mosque shootings in Christchurch, the 2021 Canterbury floods and the Covid-19 pandemic. Each of these events effectively re-energised the organisation, prompting it to 'scale up' into specific forms of action. On each occasion, there was consideration of what appropriate engagement might entail, rather than assuming that student volunteer involvement was needed, wanted or useful. The Student Volunteer Army also inspired similar volunteering initiatives at other university campuses across New Zealand, including in Dunedin, Wellington and Auckland. Such initiatives are part of an international trend in which student volunteering is encouraged and supported at an increasing number of universities.

In keeping with Sam Johnson's self-narrated identity as a social entrepreneur, the Student Volunteer Army has developed several non-disaster-related initiatives in recent years. A visiting service for older people was launched in 2016, for example, with a focus on reducing social isolation (Spink 2016). A national framework for secondary school volunteering was created, working in partnership with a number of schools across the country. The vision was to enhance societal resilience by cultivating individuals who, in the future, would be both favourably disposed and practically prepared to assist others during periods of social and environmental disruption. This initiative exemplifies the ways in which the work of Student Volunteer Army is as much a social movement as a traditional voluntary organisation.

In the 2020 annual report, which was the organisation's tenth anniversary year, the Student Volunteer Army describes its portfolio of activities as 'changing the face of volunteering'. Volunteering in New Zealand has traditionally been associated with middle-aged adults, with the 40–49-year-old age group most likely to be involved (somewhat contrary to the perception that volunteering is predominantly an activity for the retired). Bringing more children and young people into volunteering is thus literally changing the face of who volunteers. As Johnson explained, 'If we can get primary school children committed to volunteering at this age, they'll continue to do it for the rest of their lives' (Witsey 2016). To this end, the Student Volunteer Army created structures for participation that enabled young people to 'step in' and 'step out' of volunteering opportunities in a flexible manner, rather than having to sign up for a specific time period or designated role. This flexibility helped the organisation to harness the collective enthusiasm typical of young people, especially in relation to causes they regard as meaningful and of intrinsic value.

Conclusion

In the aftermath of the earthquakes, community and voluntary organisations played a significant range of roles in the city's recovery. The patterns of activity mirror those observed in other post-disaster settings, with a blend of established and newly formed organisations engaged in activities that spanned service provision, advocacy and activism (Strandh and Eklund 2018). In Christchurch, the large and established voluntary welfare organisations, many of which are faith-based, were closely involved in the emergency response phase of the disaster, providing material and psychosocial support to hundreds of affected residents across the city. The suburban churches also responded to the needs of people in particular neighbourhoods (Parsons 2014), creating community-integrated hubs of care and compassion that were typically progressive and inclusive. Although these services and activities were seldom explicitly political, in the sense of engaging critically with the policies and practices of government recovery agencies such as CERA and the Earthquake Commission, they made a positive difference for many affected residents.

In a more overtly political manner, the Māori Recovery Network and resident-led organisations such as CanCERN and WeCAN actively engaged government agencies so as to assist people with their practical needs, insurance, house repair and land zoning problems. In different ways, these organisations enabled more democratic forms of engagement during the recovery, at least for a time. With explicit support from both CERA and the Earthquake Commission, for example, CanCERN operated as a broker or intermediary, working in a collaborative manner to transcend the sometimes antagonistic dynamics of 'us' and 'them' when residents' insurance claims had gotten stuck (CanCERN 2014). WeCAN took a more combative approach in its efforts to ensure residents' voices and experiences were heard and taken seriously. Although its approach was at times strongly confrontational, given the repeated manner in which central government agencies appeared to downplay or override local concerns during the recovery, WeCAN viewed this as appropriate and proportionate. In several instances, it was also undoubtedly effective. Looking ahead, it is not yet clear what the legacy of the spaces of dialogue and contestation fostered by CanCERN and WeCAN might be. Many of the people who championed the practices of protest, brokering and mediation on which these organisations relied are still present within the Christchurch community, however, and their collective experience would be a valuable resource if there was a need to mobilise such practices again.

The emergence of the Student Volunteer Army was an unexpected development in post-quake Christchurch. Although some government organisations initially struggled with its improvisational approach, through dialogue and recalibration a way forward was developed. The energy of thousands of motivated students was thus directed into practical tasks for the recovery, with young people enabled to participate in a fluid manner, rather than insisting on more traditional forms of membership or affiliation. This work has continued to evolve in the decade after the quakes, with activities that now have both national and international reach. In terms of progressive and inclusive place-making, it is the Student Volunteer Army, the Māori Recovery Network and resident-led initiatives such as CanCERN and WeCAN that have arguably done the most to fashion novel spaces for political dialogue and collaborative action in the city. The experience of learning how to function effectively in a disaster recovery setting with a central government inclined towards top-down 'solutions' seems particularly valuable (Carlton et al. 2021; Carlton et al. 2022; Nissen et al. 2021). Such adaptive characteristics, including experimentation and improvisation, were also evident in a group of transitional organisations that arose after the earthquakes. These are the focus of the next chapter.

Part III

The city in transition

8 From transitional activities to place-making

Introduction

The possibility of regarding the multiple earthquake tremors and aftershocks experienced by Christchurch residents as a collective event which prompted a politically transformative interruption to the status quo was introduced in Chapter 1 (Cloke, Dickinson and Tupper 2017, following ideas from Badiou 2005). In short, it is argued that such an event can serve to rupture existing place narratives by allowing the emergence of alternatives to the current order. Such alternatives often involve previously marginalised but potentially disruptive forms of imagination and action (Dewsbury 2007), which can foster a wider appreciation of and confidence in (or 'fidelity' to) new ways of knowing about, and being in the city. In turn, these new narratives of the city can offer new forms of visionary cultural politics and ethics that add complexity to analytical narratives that suggest a simple recalibration of previous neoliberalised governance after the event of the earthquakes. In the Christchurch context, Pickles (2016) has argued that alongside the catastrophic damage to the geo-spatial surface of the city, the impact of the earthquakes prompted a less visible but equally significant rupturing of the ways in which it is underpinned culturally by narratives and practices of place.

The previous chapter showed how the voluntary sector played a significant role in nurturing new cultural-political values, and ethical spaces and practices in amongst the rupture. This chapter focuses on a particular movement within the voluntary sector that was responsible for a series of 'transitional activities' in Christchurch which marked not only a temporal stage in the rebuilding of the city but also a potential shift in how its collective life could be re-imagined and performed. Following arguments developed by Cloke and Conradson (2018) and Cloke and Dickinson (2019), it can be suggested that the cultural 'rupture' identified by Pickles was characterised at least in part by the emergence of previously hidden and little-noticed ethical and aesthetic ideas and practices that came out of what might be thought of as transitional forms of urbanism. Such ideas and practices became recognisable in the material and symbolic artistic expressions and creativity that sprang up in temporary activity spaces that became available in demolished areas of the city. Although these activities were sometimes assumed to be short-term

DOI: 10.4324/9780429275562-11

gap-fillers, destined to be overtaken by formally planned rebuilding, for visionary leaders of the transitional movement in Christchurch they offered the opportunity to weave ethically progressive ideas and material practices into the re-emerging fabric of the city. Their collective work comprised an innovative series of diverse and nimble community-focused interventions that drew heavily on participatory arts and ecological action in ways that brought an ongoing sense of being in transition to the longer-term evolution of the city. They also attracted international attention, exemplified by recognition for Christchurch in the 2012 *Lonely Planet* travel guide as a top-ten tourist destination where energy and inventiveness were exuded from the pop-up venues and inventive artworks that graced empty demolition sites (Wright 2012).

A cultural politics of place

The impact of these transitional activities needs to be set within the context of key characteristics that shaped the cultural narratives of Christchurch prior to the earthquakes. Three principal narrative traits have been identified. First, the city traded on a quaint and tranquil version of quintessential Englishness (Cupples and Glynn 2009; Schöllmann et al. 2000), formed both from an erasure of pre-colonial histories and from a deliberate emphasis on cultures and styles of Anglophilia. From built environments and place names to educational and religious rituals, the 'garden city' of Christchurch was developed and maintained in tune with the home-based values of its predominantly English settlers. However, according to Pawson (1999), emerging changes with respect to land rights and commodification of property constituted an alienation from previous assumptions about appropriate vegetation and land use, resulting in an ambivalent relationship between colonial imagery and indigenous environments in the more-than-human city prior to the earthquakes (see Chapter 11).

Secondly, national-level fascination with neoliberal experimentation since the 1980s had helped to shape more local narratives of the political and economic character of Christchurch. Hayward (2012) concludes that since then the planning of the city has been characterised by a largely permissive regime that has given entrepreneurs a relatively free reign, with elected local authorities reduced to a role of mitigating negative impacts of investment rather than a more pro-active and participatory shaping of the city (see also Chapters 4 and 5). To some extent, then, the resultant localised expression of neoliberal management cultures in business and land ownership tended to result in influential endorsement of central government policies that were so advantageous to them. However, it should be emphasised that alongside these conservative traits, there were elements of practice and forms of organisation in the city that worked against top-down neoliberalism, helping to form an accompanying cultural imagination that the city also functioned as an inclusive democracy, with a progressive left-of-centre social politics (Marcetic 2017; Pawson 2022).

Thirdly, the role of the arts in Christchurch prior to the earthquakes was dictated largely (but not exclusively) by commercial pressures and dominant artistic tropes favouring 'art for art's sake' rather than the development of artistic spaces of political potential or social transformation (Newman-Storen and Reynolds 2013). By and large, the arts might be considered as a somewhat docile presence in the pre-earthquake city, pushing against local conservatism from a relatively peripheral position (Parker 2014), generally not disturbing dominant place narratives and struggling to gain traction as a platform for radical alternatives to the neoliberalised structures and cultures of the day.

Of these three narratives, it is often assumed that it is the dominance of neoliberal political economy which has been least affected by the earthquakes. In this interpretation, as outlined by Hayward and Cretney (2015), the post-disaster planning environment was shaped initially by the imposition of a centralised command-and-control model of recovery that consisted of building demolition and government acquisition of land that was then used for anchor projects or sold off to private investors to support market-led projects (Dann 2014; Dickinson 2018). In reality, the situation was more complex. While democratic scrutiny and public participation were not as great as some hoped for, subsequent redevelopment in the city centre – as shown in Chapters 4 and 6 – was based on a combination of local and central government planning, Crown land purchase, mandatory privately-held land title amalgamation, market-led and often insurance-funded interventions and private and public sector place-making. It was this combination of approaches that underpinned ideas about success and hope for the future of the downtown's built environment.

The dominant historical narratives of the city were however challenged in other ways by the rupture, which offered opportunities for different kinds of understandings and activities to influence the identity of the city. Rebecca Solnit (2009) has demonstrated that disasters can provide space for freeing up community potential and addressing community goals that find expression as much in the interstices of social reconnection and shared action as in the materiality of construction. One distinct example of such release in the Christchurch context has been the provision in post-quake legislation for the incorporation and expression of Māori design and values within the city's redevelopment (see Chapters 4 and 9). This has helped to ensure the rebuilding of the city is occurring in a manner that more directly honours the partnership seen as central to the Treaty of Waitangi. In particular, the work of the Matapopore Charitable Trust has championed the essential task of embracing the voice of mana whenua – the status held by Ngāi Tūāhuriri, the local Māori subtribe with territorial authority – in different aspects of the city's recovery.

Matapopore's activities have ensured that Ngāi Tūāhuriri and Ngāi Tahu values, aspirations and narratives have been woven into redeveloped urban environments, especially through the use of Maori design references and motifs (Figure 8.1). There is now more evidence that the previous marginalisation and exclusion of Māori values and identities within the Christchurch

Figure 8.1 Examples of Matapopore interventions in the new urban design of Christchurch: (top) one of 17 whāriki – Ngāi Tahu welcome mats integrated into riverside walkways; and (bottom) artwork representing a Māori cloak of kakapo feathers, disguising the car park at the Justice and Emergency Services Building.

built environment is being addressed in the aftermath of the earthquakes (Ballard et al. 2015). Some recovery decisions have clearly upheld elements of previous place narratives – for example, in a move widely attributed to a reaffirmation of European settler heritage values and identities, the colonial nineteenth-century Anglican cathedral is being rebuilt in its original form and central position in the city centre. However, the renewed built landscape of the city does suggest that some historical expressions of Englishness are beginning to give way to a more overtly diverse urban landscape, reflecting the presence of both Treaty partners, iwi (the tribes) and the Crown, and the city's wider culturally mixed population.

The rest of this chapter focuses in on the changing public role and function of the arts in post-disaster Christchurch, and on how particular forms of ethical creativity have emerged in the material spaces created by the earthquakes. It traces the emergence of new forms of non-governmental organisation after the earthquakes, the activities of which represent energetic and multifaceted contributions to cultural, artistic, creative and ecological uses of temporary spaces and to the wider aesthetics and ethics of transition. The status and contribution of these 'transitional' organisations is however contested; they have sometimes been portrayed as whimsical and colourful band aids (Weejes Sabella 2015) with no meaningful role in longer-term recovery from disaster (Wesener 2015). Alternatively it can be argued that such agencies have demonstrated a much more significant potential as agenda-setters in an experimental search for new ways of inhabiting and organising public space. From this perspective, they articulate new understandings of political and ethical citizenship that embrace and encourage a role for residents as active participants in new forms of urban place-making.

Developing and performing the transitional city

Transitional organisations emerged with some rapidity in the years immediately after the earthquakes. Informal and in some ways latent networks became activated by key individuals in calls to collaborate in new volunteer-based ventures to spark new forms of cultural diversity and biodiversity around the city. It is important to note from the outset that these transitional concerns, and the subsequent energy to address them, were socially and spatially variegated, with three factors exerting particular influence on the patterns of activity. First, the lifestyle priorities and coping strategies of different suburban populations were significant, with some clusters of suburban population demonstrating low levels of affinity for, or compatibility with transitional activity. For example, it has been suggested (Dombroski 2020) that after the initial period of emergency response, during which close networks of community aid and support were active, many long-term residents of the city may have been relatively overwhelmed by the negative effects of the earthquakes, either by damage to property or by impact on place identity or both. Understandably, often people became focused on returning to a personal sense of suburban normality rather than taking a civic-level or even community-level view of the post-disaster landscape. Their relative disinterest in city centre projects may also have been exacerbated by previous periods of lifestyle change in which the suburbanisation of city life had for many reduced the usefulness of and familiarity with the central city.

Secondly, certain suburbs such as Lyttelton, New Brighton and Woolston demonstrated a greater affinity with the creative responses of transition. These were places where demolition sites provided opportunity for experimentation, but they were also places with a strong history of community and artistic group activity (Lesniak 2016) where latent networks were more easily sparked into life by transitional activists. Thirdly, Bennett (2016) has identified a significant group of relatively young, more artistic and entrepreneurial people for whom the central city of Christchurch had become a vibrant experimental location rather than an outmoded downtown space. He comments that

> I think that going back to the five years or so before the quakes that had changed quite a lot, and for a minority, but I think for a significant minority, which I think is most of the same people involved in the transitional movement, central Christchurch had become a really exciting place.
>
> (Bennett 2016, 154)

Although the scale of such activism may be debatable, this newfound appreciation of the central city by key activists contributed significantly to the transitional movement that emerged after the earthquakes.

Much of the initial energy for transitional activities came from dynamic younger people in these existing artistic and entrepreneurial communities in the city, but kindred spirits from elsewhere also became motivated to bring their creative ideas into the post-earthquake context (Boswell 2021; Lesniak 2016). Forming significant new groups and connections, these innovators spotted and acted upon the opportunities presented by the earthquakes to express new kinds of transitional place-identity. In some ways, these were people from the cultural and political margins. Their existing expertise in art, theatre, design and landscape architecture was ready and waiting when the post-earthquake rupture presented them with fertile opportunities to practice their arts in new ways. As a consequence, evaluation of the impact of transitional organisations will rely as much on their capacity to create new ethical and aesthetic expressions (or 'vibes') in the city and to open out possibilities for new community-centred projects and economies, as on the numbers of residents participating in or approving of the activities concerned.

The rupture of the earthquakes provided a physical and cultural landscape in which art, design and informal theatre had a new place and role in the remaking of the city. Individually and collectively, artists, designers and ecologists began to make multiple contributions of various forms, including large-scale murals and temporary art installations, typically sponsored by business interests, which provided aesthetic contrasts with and challenges to both the emptiness and abandonment of the previous cityscape and the slow development of its glass and concrete successor (Figure 8.2). It is noticeable

Figure 8.2 Street art in Christchurch.

that amongst the plethora of mural images – some of which simply emit a cheerful sense of fun – there were also emergent underlying themes related to the importance of Māori culture and ecological sustainability in an age of multiple disasters.

Perhaps the greatest impact on the rethinking of the city, however, was the activation of vacant spaces in the city by a series of transitional voluntary-sector organisations formed in the immediate aftermath of the earthquakes, notably Gap Filler (founded in 2010), Greening the Rubble (also in 2010), Life in Vacant Spaces (in 2012), and the Festival of Transitional Architecture (or FESTA, in 2012). Together with some support from the city council, these organisations curated an experimental creative movement in Christchurch, using a series of temporary installations and performative practices to reclaim and recode urban spaces, and in so doing illustrating the potential for a longer-lasting place for transitional creativity in the city. The resultant creative surge was recognisably catalysed by the event of the earthquakes – not only in terms of the materiality of suddenly vacant spaces, but also in a new ground-level cultural politics of doing what is necessary without waiting for more official responses.

Gap Filler is a creative placemaking agency operating at the crossroads of community development, urban design, art and public intervention. In the words of Ryan Reynolds, its co-founder and creative driving force:

'We've been conceiving events and installations, building community and experimenting amidst the ruins of the nineteenth and twentieth century city. We temporarily reshape the urban environment in a quick and rough way, without waiting for the government's or developers' plans and permissions. It was predictable that people here would take what they could into their own hands. When the normal ways of doing things are inadequate, those in power have to start revising the rulebook and reasserting control. Others inevitably start doing what they feel is necessary and lacking, and hope that the rulebooks will catch up'.

(Reynolds 2014, 167)

The resultant sites and practices of transitional ingenuity have already received considerable attention (see Bennett et al. 2014a, 2014b; Carlton and Vallance 2017; Cloke and Conradson 2018; Cloke and Dickinson 2019; Cretney and Bond 2014; Dionisio and Pawson 2016; Dombroski et al. 2019; MacPherson 2016; Parker 2014; Wesener 2015).

It is however worth noting that each of the main organisations has brought particular perspectives to the collective movement. Gap Filler has championed an openness to ingenuity and innovation, using participatory propositions as a catalyst for alternative performances of the city. As Boswell (2021) indicates, this propositional approach represents an experimental provocation with which to test the water for change. It begins with a conversation that is not already present in the city. It then trials new ideas before putting a particular project 'out there' to assess levels of public engagement. An inherent quality of such experimentation is that it accepts failure as a potential out-come. Projects that have attracted public engagement include a temporary pallet pavilion providing a stage for events, a washing machine repurposed as a coin-operated jukebox and twinned with a sprung wooden dance floor to create the public Dance-O-Mat, and a giant outdoor arcade game system, the Super Street Arcade. Greening the Rubble has been responsible for a series of temporary community gardens and memorials throughout the city, often using mosaics of materials and recycled plants from earthquake affected areas (Figure 8.3). Other organisations have been similarly experimental. Life in Vacant Spaces has developed a specific facilitation role with landowners and businesses and community groups, fostering ecological and entrepreneurial innovation in a range of spaces. FESTA has curated a series of annual weekend events as well as more regular creative conversations designed to provoke a re-imagination of design and architecture in the city. Thousands of people attended its annual events held on cleared sites in the city centre once the cordon receded.

These initiatives have been evaluated with varying degrees of enthusiasm. For participant organisations and individuals, transitional initiatives have encouraged 'risk-taking, discovery, a not knowing how things will turn out, an openness to change' (Winn 2015), but are also understood as far more than temporary stop-gaps to be overwritten by the more formal and permanent architectures of the new city. Although the projects themselves may be short-lived, the process that underlies them is designed to endure beyond the timescale of the temporary. It encourages a participatory and experimental stance of being prepared to try something new, seeing what works and what doesn't, and through this experience gaining valuable ideas about what kinds of alternative futures can be brought into being. These early years of transition, then, make a claim on the future city through experimenting with and embodying the use of temporary urban sites. Parker's (2014) assessment of transitional projects emphasises how they bring people together in experimental ways in particular spaces, and in so doing open out new capacities for acting and connecting differently with the city. By creating space for residents to perform the city in new and creative ways, transitional organisations have modelled a form of participatory place-making that is able to better engage many citizens in processes and practices of involvement in the contexts in which they live.

On the other hand, transitional initiatives have also been the subject of negative commentary. The aesthetics deployed have been dismissed as 'messy' and 'bohemian' and therefore out of character with the historical character of the city (Lynch 2014). The scale of the projects is regarded by some as too small and ephemeral to make any lasting contribution. Participants are thought to be already-existing community activists, and beneficiaries are as likely to be tourists as city residents. A third line of critique argues that in and through the financial support received from Christchurch City Council, transitional organisations have inadvertently been co-opted as showcases of the state's short-term entrepreneurial responses to the earthquakes. In this way,

Figure 8.3 Examples of transitional projects: (top) Gap Filler's Pallet Pavilion in
2013; and (bottom) Greening the Rubbles' Green Room in 2015.

transitional activities are dismissed as small-scale entertainment that lack the means to impact on bigger picture problems (Grimshaw 2015).

Two important questions aid analytical navigation between these potentially polarised perspectives. First, what does a deeper examination of transitional organisations and their activities reveal about the values and aesthetics they deploy, and their capacity to engender 'affective communities' – that is, forms of community distinguished by widely held and collectively understood forms of feeling? Secondly, with a decade having passed since the establishment of transitional organisations in Christchurch, how does their evolution into mature actors in the city reflect longer-term success in sustaining these objectives in amongst the terrains and politics of the rebuilding of the post-disaster city?

Transitional ethics, aesthetics and affective communities

The leaders of transitional organisations in Christchurch have consistently argued that the curation of transitional spaces of ingenuity and experimentation does not equate to a formal political process. Rather, these activities are seen as underpinned by a series of ethical values and a desire for social impact. Using the example of Gap Filler, Cloke and Dickinson's (2019) analysis identified three strands of the underlying values that were being introduced into the city through the performances in different activity spaces.

The first is that participatory practices inherent in many Gap Filler experiments helped to build an imagination of the city that was able to convert a pragmatic and short-term view of the temporary into a longer-term and more visionary understanding of a sustainable role for experimental place-making. This spilled over into some commercially focused developments, such as the container mall, and later, initiatives like Little High and Riverside Market discussed in Chapter 9. So, in practice, the rupture of the earthquakes was seen to create both material and imagined spaces in which the rethinking of the city could occur. The gathering of people in short-term vacant sites enabled those who otherwise felt excluded from the rebuild process to join in with informal and often unstructured practices and performances that counterposed the blueprint being imposed by central powers. In these activity spaces, there was an unambiguous permission to experiment, to take ownership (however small) of an emerging art of the possible, and of the acting out of potential. In so doing, individual participation projected an imagination of a Christchurch which could be more collaborative and inclusive (Reynolds 2014, 169). Transitional activities therefore engendered an agile form of social entrepreneurship that suggested the need to embed ideas and practices of experimental bottom-up city-building into future phases of urban regeneration.

The second set of values detected in the work of transitional agencies reflects an inherent desire to propagate ways of living and working together that enable acceptance of and adherence to particular kinds of civic values that replace previous conservatism with more relaxed, interesting and open-minded forms of political culture. The top-down implementation of the CERA blueprint for the city (discussed in Chapters 4 and 6) strengthened perceptions in some quarters that only a limited elite would have a say about the future of the city. Alternative transitional activities emerged at least in part in response to that imagined elitism, and deliberately sought to enable a more common involvement in the recovery not least by identifying civic rights as fundamental to their activities. These values are potently illustrated in the development of the site of the demolished Crowne Plaza Hotel, which had been built in the 1980s across part of a public street closed for this purpose. After 2012, it was returned to the people as 'The Commons', becoming the base for Christchurch's community of transitional activists (Ballard et al. 2015) and the site of the Pallet Pavilion (Figure 8.3). Users of this gathering space are subject to guidelines which reverse the normal rules of ownership, ensuring that projects are focused on community engagement, related to the post-earthquake environment, engaged in social change, interested in collaboration, make space for local creatives and demonstrate a pragmatic fit to the site's transitional ethos. This mix reflects a deliberate attempt to incubate forms of in-commonness which are collective, collaborative and widely inclusive, offering open opportunities for creative place-making.

Thirdly, the activities of transitional organisations in Christchurch made deliberate use of particular aesthetics to suggest the possibility of ethical action. The term 'aesthetic connection' originated from key thinkers in art and social sculpture, especially Suzy Gablik (1992, 1996) and Joseph Beuys (Beuys and Harlon 2004), who argued that particular artistic and theatrical aesthetics can help to stimulate people's imaginations and form subtle but significant connections with particular social and environmental values. The performance of transitional activities in Christchurch made use of these kinds of connections with the transgressive potential of the post-earthquake rupture. Whether by storytelling about possible futures, or bringing together material, performative and symbolic elements to suggest wacky expressions of place-making possibilities that divert from previous norms, or by using quirky and humorous aesthetics to allow residents to discover new directions and meanings counterposed with the formal rebuilding of the city, aesthetic connections were used to draw residents and communities into the enactment of re-making place after the earthquakes. It can be argued that this mix of aesthetics and values represented a re-staging of the city, enabling a re-sensitising of memory and a re-envisioning of futures via creative participation.

The deployment of these three strands of values by transitional organisations has involved many different events, supported by thousands of volunteers working with hundreds of different businesses and community groups (Reynolds 2015) (Figure 8.4). However, the contribution of transitional organisations to the widening participation of place-making in Christchurch may be more far-reaching and complex than simple numerical expressions of awareness and participation might suggest. Cloke and Conradson (2018) note that the curation of sites and spaces of shared action, social reconnection

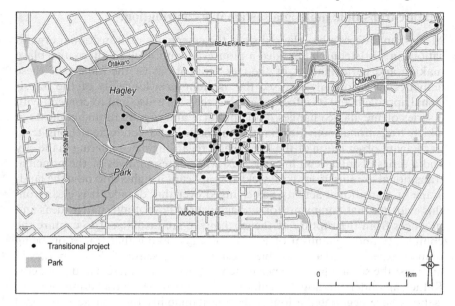

Figure 8.4 The distribution of transitional activity sites in central Christchurch.

Source: 'Pop up projects' at https://findchch.com/categories/whats-hot?subcategories=transitional-city-projects&page=all

and aesthetically connected energy became associated with emotional responses that gained traction in the city, creating a collectively experienced set of feelings, most often narrated as a 'buzz', a 'vibe' or a 'new atmosphere' (Campagnolo 2015; Merritt 2015; Winn 2015). Likewise, Cretney (2016) equates transitional activities with the opening up of new avenues for social participation, presenting geographically dispersed communities with a set of overlapping opportunities, and fostering a palpable atmosphere of hopefulness in amongst the landscapes of urban disaster. MacPherson (2016) refers to the role of transitional organisations in the creation of a new 'spirit' in the city.

The importance of these traces of emotions should not be underestimated when assessing the significance of transitional activity in the city. Even if the numbers of participants in specific projects is a small fraction of its citizenry, the creative alterity that they fostered became evident more generally as part of an emerging narrative of post-earthquake Christchurch as a city of cultural innovation and novelty. The early activities of transitional organisations brought together shared patterns of emotional understanding and meaning which lent some coherence to aspects of local community. Although these activities avoided direct forms of political participation, they nevertheless laid the foundation for an affective politics of transitional urbanism in the post-earthquake city. This has also been carried across into a lasting embrace of street art and some interesting commercial experiments in place-making explored in the next chapter.

Mature transition: making a sustainable contribution?

More than a decade has elapsed since the emergence of transitional organisations in Christchurch heralded a pioneering approach to alternative place making and performances of the city. Evaluations of the impact of these activities have not only assessed the medium-term longevity of the transitional approach and its effectiveness, but they have also had to contend with a series of inevitable changes in the political and organisational contexts in which transitional activities have been shaped and conducted. Long-term participant observation of the key debates regarding place-making in the city and a series of interviews in early 2020 with representatives of the principal transitional organisations have yielded significant insights into how the landscape of transitional activity has changed over that decade. It is clear from this evidence that arguments for and against the flow of alternative ideas and practices from the rupture of the earthquake are insufficient to capture the complexity and dynamism of a form of city-making that trades on being nimble, experimental and capable of establishing a sense of the temporary in amongst the seemingly more permanent rebuild of a concrete and glass city. Rather, an understanding is required of the ways in which transitional organisations have negotiated a longer-term sustainability for their activities and of how responses to the changing conditions of the city have altered the capacity to maintain a fidelity to the ethical values and aesthetic connections that characterised early forms of transition after the earthquakes.

The changing nature of political governance and planning of the post-disaster city of Christchurch has been examined in detail in Chapter 4. To some extent, there has been a shift in the obvious drivers of the previous dominance of top-down politics and governance in the city. The post-earthquake regime established by CERA has been replaced by a new emphasis on decision-making by the city council and its mayor between 2013 and 2022, Lianne Dalziel. However, the central state remains influential in the capacity of local government to make significant changes of direction to previously dominant market-led agendas. Since 2017, Labour-led and Labour governments have shown evidence of being open to new ideas about experimental urbanism, expanding earlier nationwide support of city cycleways for example. But they have if anything strengthened a significant instinct for centrally effected change (for instance in abolition of regional health boards and water management) that restricts the agency of local forms of governance.

It is telling that central government has been regarded as unable – despite its endorsement of ideas about experimental urbanism – to raise levels of recognition and funding for non-profit organisations such as those at the heart of experimental urbanism in Christchurch (Vallance 2020). The mayor and city council have always been viewed as broadly sympathetic to grassroots initiatives and have continued to provide some funds for the work of transitional organisations as part of their attempt to re-engage the hearts and minds of people in the city with more localised politics of urban renewal. Such support is significant, but inevitably limited by financial capacity.

Moreover, an emerging culture of 'metrics' to ensure both that expenditure is accountable, popular and manifestly beneficial to residents, and the observance of often minor but compelling health and safety regulations affecting activities across the city, has established a decision-making environment that to some extent clashes with the experimental and risk-taking nature of many transitional activities (Welfare 2020).

This shifting political context has coincided with and helped to shape, the evolving organisational context of transitional activity in Christchurch. First, it is evident that discourses of 'transition' are being replaced by those of 'place-making'. Place-making may be broadly understood as a process that inspires people to engage collectively in the reimagination of the public spaces at the heart of their community, thus helping to co-create contexts where they live or can see their future (Ermacora and Bullivant 2016). To some extent this discursive shift has been led by the city council, whose initial funding for transitional activity was organised under the Transitional Cities Programme to support post-quake recovery and well-being. This was later rebranded as the Enliven Places Programme, which aimed to transform underutilised sites into vibrant and welcoming public spaces (Christchurch City Council 2022a). The new programme prioritised fresh discourses; support was still offered to the community to deliver temporary projects to enhance amenity and activity, but such projects were encouraged to deliver a lasting legacy to the identity and sense of place in the city. This political lead, which shaped council funding for the activities of non-profits, helped to establish a new discursive environment for the organisations concerned. The idea of transition remained meaningful for some community partners and artist communities (Dombroski 2020) who recognised that the transitional phase after the earthquakes was by no means completed, but political expediency and organisational necessity led to a strongly-mediated recognition of place-making as the dominant theme for longer-term participation in the city. The discursive connotations of 'transitional' activities as a form of slapdash and unfinished do-it-yourself response to the earthquakes (Welfare 2020), were replaced by more slick and seemingly strategic connotations of place-making.

Alongside these discursive manoeuvres, the voluntary-sector organisations concerned underwent different kinds of changes in search of longer-term sustainability. Gap Filler, for example, transformed itself into a place-making social enterprise, using its post-earthquake reputation to offer design and delivery strategies for creative and experimental civic installations for communities and the public and private sectors in Christchurch, other New Zealand cities and internationally. A prime example of this new approach is a partnership with building company Fletcher Living to help create central city residential communities in the east frame of the central city (Gap Filler 2022). Drawing on its experience of using temporary activations to help foster longer-term community outcomes, Gap Filler is facilitating a range of community-minded projects as well as filling in the gaps with its own style of projects. This more formal integration into developer-led placemaking permits not only more sustainable sources of funding for the organisation, but also an opportunity to ensure a flow of playful and participatory aesthetics into longer term residential community-building.

Greening the Rubble has also sought to transform itself from an earthquake-response organisation into a more sustainable social enterprise. In 2019 it was rebranded as The Green Lab (2022), offering customised services for indoor greening projects and outdoor landscaping for public and commercial areas. Again, working as a social enterprise permits the organisation to maintain its transitional commitment to supporting community well-being and connecting people to nature. It also accommodates an adaptation to the evolving needs of the wider city via creation of community-led green spaces in areas of residential regeneration. Its new approach is illustrated by the Linwood Tiny Shops project, which in partnership with Te Whare Roimata and the city council involved the construction of a micro-village of small buildings in a suburban community garden, to be used as shops and a café and to support a range of local events.

Rebranding was also a strategic decision for the Festival of Transitional Architecture. Now renamed as Te Pūtahi, it organises Open Christchurch – a weekend festival that began in 2021 for residents to experience both traditional architecture and post-earthquake design in the city – whilst also hosting a diverse range of events to encourage civic awareness and engagement (Dionisio 2020). These events include gatherings, such as Christchurch Conversations, that connect Christchurch professionals and community groups with a range of international place-making concepts and networks, and showpiece projects such as the Old School House (Te Kura Tawhito) in the coastal suburb of New Brighton. Here architectural design expertise supported local community groups in the sympathetic adaptation of the building as a location for a collective of creatives, craftspeople and local organisations (Te Pūtahi 2022).

Perhaps the main exception to these adaptive trends is demonstrated by decisions taken by Life in Vacant Spaces to continue its charitable and community-centred work without rebranding or adopting socially entrepreneurial strategies of property and land management. Its early focus on performative pop-up projects in disused outdoor spaces in the inner city has evolved into brokering the development of pop-up shops and indoor art exhibitions in the internal spaces offered by vacant retail and office sites. Moreover, its commitment to helping local communities to help themselves has required a stronger focus on brokering entrepreneurial activity and community self-help in more suburban locations, and in the residential red zone corridor. For example, the East x East project involves a significant area of red zone land in Burwood that Life in Vacant Spaces (2022) has licensed from Land Information New Zealand, the Crown agency that has held ownership of that red zone land in Christchurch where housing has been demolished because of earthquake damage (Chapter 12). The objective for East x East was to create a vibrant and community-orientated space filled with temporary art projects, events and activities that demonstrate a commitment to communing, by listening to local communities seeking sites and practices where in-commonness can be pursued.

In each of these examples, organisations that arose from the rupture of the earthquakes have variously sought to evolve from earthquake-responders to sustainable agencies of place-making, often recognising the necessity to work together in partnership with government funders and commercial interests as well as local community groups. Although in each case, organisations have aimed to carry their initial experimental creativity and underlying senses of ethics and aesthetics through into these mature phases of social enterprise and partnership, they will inevitably be open to critiques involving selling-out, loss of ethical drive or sanitised incorporation into the agendas of more powerful partners. In conclusion, this chapter turns to the two main evaluative tropes which have been used to gauge the longer-term contribution of formerly-transitional non-profits in Christchurch.

Conclusion

The first of these tropes mourns the loss of significant opportunities that were presented by the event of the earthquakes. The initial post-quake period is seen by many (and certainly by those directly involved) to have been characterised by an outburst of positive energy for change in the city. This was recognisable both in an opportunity for community involvement in the priorities for, and direction to be taken by, renewal in the city, and in the stream of creative thinking and experimental activity exemplified by transitional organisations. This creative energy provided a focus for transformational change, involving a capacity to deviate from the past and to experiment with aesthetics and ethical values which enabled a rather different vision for narrating and participating in the life of the city. It gave rise to hope and even expectation that wider redevelopment processes in the city could be fuelled by creative thinking and collaboration between the public and private sectors and the key non-profit organisations responsible for early experimentation (Fisher 2020). In this way the ethical values and aesthetics of the transitional organisations could have been fostered significantly in the subsequent redesign of the city, in particular by adopting processes involving meaningful liaison with local communities and eschewing formulaic space-filling in favour of more experimental and risk-tolerant project design and decision-making.

In the political context that prevailed after the earthquakes, CERA's top-down control of the rebuild process gave little acknowledgement of, or support to, the work of key transitional organisations, and the blueprint for planning and development of anchor projects in the city largely sought to provide functional enhancements for what had existed previously. There were exceptions to this generalisation, such as the Margaret Mahy playground discussed in the next chapter, and the more open-ended processes of residential red zone planning explored in Chapter 12. With a political desire to support existing landowners and developers, the opportunity to foster transformational place-making projects was overlooked. With the significant exception of commercially sponsored street art, the result for some commentators has therefore been a rebuild process of business as usual, resulting in a city characterised by 're-creation of the same old same old, with a few bits of public art thrown in' (Fisher 2020), and the loss of any serious engagement with the energy and creativity of transitional organisations.

The second trope offers a more optimistic perspective, recognising that the early post-quake freedom for transitional activity unleashed an enthusiasm for collective and inclusive ways of working that has resulted in an enduring sense of alternative possibilities for place-making in the city. Accordingly, there is a hopefulness that the genie of creative aesthetic connection and experimental participatory action has produced lasting legacies. For example, the new approaches and aesthetics arising from partnership with organisations and individuals representing Ngāi Tahu and Ngāi Tūāhuriri have resulted in a more-than-tokenistic weaving of the values, aspirations and narratives of Māori people into the fabric of the post-earthquake city and some of its buildings. This represents a significant break from earlier narratives and encourages a hopeful expectation that other expressions of the cultural politics of local community might endure.

If over time the capacity for transitional organisations to make a sustained impact on place-making has been squeezed, there are also signs that the thinking behind transitional activities remains active in certain spheres of the city. New spaces of involvement have continued to open up, for example the shift to indoor sites (such as the Imagination Station for children occupying a floor of the new central library, Tūranga), the partnership between Gap Filler and Fletcher Living as part of new residential development in the east frame, and the long term opportunities for a second flush of transitional experimentation with new forms of place-making in the city's red zone corridor. Such reorientations indicate a sense in which there will always be elements of flux in urban development and redevelopment in which the creative energy of aesthetic connection and ethically in-common participatory community activity can take root.

Both evaluative tropes – telling stories of lost opportunity and of enduring hopefulness – pose important questions about the temporality or otherwise of the transitional activities that emerged from the rupture of the earthquakes. Transitional challenges to the orthodoxies of public space have demonstrated the new kinds of cultural politics that can emerge when different elements of the community at large become involved in the evolution and making of the city. As other potential disasters loom large, not least those associated with the regulation of carbon footprints and the urgent necessity for more sustainable resource use, so the affective politics inherent in transitional thinking remain a paramount rejoinder to business-as-usual practices.

9 Landscapes of consumption

Introduction

One of the striking effects of the earthquakes was the sudden loss of city centre and suburban sites of consumption associated with leisure, retailing and culture. Well-loved restaurants, cafés, bars, shops, public facilities such as cinemas, pools and libraries, and the heritage buildings that gave the city a sense of place: these disappeared overnight. Many were damaged beyond repair, lying isolated within the city centre cordon (Figure 6.2). As the cordon shrank, and life slowly returned to that broken landscape, individual events seemed like big milestones. Ballantynes department store, a fairly modern low-rise building, reopened in late 2011, becoming the anchor for the container mall. A year later, C1 Espresso café re-located to a solidly re-purposed old post office, quickly assuming the role of meeting place and downtown salon: there was little else. The first restored cultural facility, the Isaac Theatre Royal, opened in November 2014. In April 2016, on Anzac Day, huge numbers of people gathered when the Bridge of Remembrance – a first world war memorial that now functions as an entry point to the city's pedestrian mall – reopened after repair.

This chapter considers the role of consumption in post-earthquake recovery, and both the continuities and transformations from its pre-earthquake forms. Consumption in this sense denotes intertwined experiences, sites and products. It is much more than shopping, most of which anyway occurs in suburban malls; indeed the blueprint recovery plan considerably diminished the area of the central city set aside for this function. Instead the chapter conceptualises the city and its centre as a focus of consumption flows, competition for which is between places, at both the intra- and inter-urban scales. Consumers are local, national and in post-earthquake planning (but not necessarily post-Covid recovery) international. Large anchor projects such as Te Pae – the convention centre, the as-yet unfinished 'multi-purpose arena', the metro sports centre, and the very successful Margaret Mahy Family Playground were designed to attract such consumers but, as Chapter 6 demonstrates, to hold property investors and their capital in place as well. Consideration is

DOI: 10.4324/9780429275562-12

given in turn to retailing and hospitality; to the anchor projects purposed to support events and tourism; and to cultural heritage projects and restorations. It concludes by reflecting on how the spatial competition between places to attract consumers and consumption practices depends on and drives a further cycle of consumption, that of carbon.

Creating landscapes of consumption

The global rise of consumption and the commodification of leisure have been central to contemporary capitalist growth (Zukin 1998). This phenomenon is underpinned by a work-and-spend cycle, stimulated by advertising to fuel desires to own and engage with an ever-expanding range of commercially provided things, services and practices (Perkins and Thorns 2011). In this consumption process, individual and collective identities are constructed through the purchase and possession of objects and associated experiences, and performed in sites of 'aestheticised entertainment and pleasure' (Meethan 1996, 324), known also as landscapes of consumption. In Christchurch, in the two decades before the earthquakes, planners had therefore sought to emulate international trends in the creation of such landscapes. The city centre's streetscape was described as the product of 'waves of interest and investment in central city functions, the most recent of which [was] based on leisure and tourism' (Pawson and Swaffield 1998, 254). The focus on encouraging consumption complemented but contrasted with an earlier municipal drive – which remains strong – to provide facilities such as libraries and galleries, parks and playing fields. The city retains over 1200 public parks and gardens.

Landscapes of consumption are created in practices of place-making and promotion, which in the case of Christchurch have long built on its reputation as a fairly traditional 'garden city' (Pawson 1999). Since the earthquakes, as the previous chapter discussed, place-making techniques have been used creatively as part of the on-the-ground process of rendering the static geometry of the blueprint into what Jan Gehl would call 'people cities' (Matan and Newman 2016). They help to adapt the efficiency that characterises modernist planning into places for human opportunity, places that draw in people and encourage them to linger. An important part of this is making people and their interests (rather than traffic) visible. In the central city, which is renowned for its windiness, this has translated into the provision of sheltered laneways and squares, and a new emphasis on the riverside precinct, Te Papa Ōtākaro, as an attraction and locus of rest and recreation in its own right. This precinct is one of the more popular anchor projects, rooted as it is in the city's garden image of itself. The stories of mana whenua are woven into its design (Figure 8.1), in recognition of a more contemporary understanding of garden landscapes (Landscape Architecture Aotearoa 2022; O'Callaghan 2021).

The blueprint's focus on the central city has to be understood alongside the rise of suburban malls and the decline of central shopping streets. Globally this has been a major trend for decades (Teller 2008) and as part of the rebuild suburban libraries, sports facilities and other sites of cultural belonging have been rebuilt or replaced. In the central city, the night-time economy has been a notable element of Christchurch place-making prior to and after the earthquakes (McDonagh 2017), as discussed in Chapter 6. Internationally, this phenomenon has its basis in 1970s attempts to revalorise de-industralised inner cities to capture both real estate value and cultural capital (Lovatt and O'Connor 1995). Such investments aim to boost adult consumption and leisure experiences by adaptively repurposing redundant or under-utilised buildings for niche retailing and hospitality, but with a particular focus on bars and nightclubs. While these facilities and inner city landscapes have attracted consumers to replace those lost to the suburbs, they have also been associated with unhealthy drinking practices, especially among young people (Roberts 2006; Shaw 2010). Inner city Christchurch before 2010 was no exception (Johnston 2014; McCrone 2012b).

Place-making in Christchurch since the earthquakes also recognised that tourism has been perhaps the ultimate consumption industry globally, and before the Covid-19 pandemic was New Zealand's largest export sector (Perkins and Rosin 2018). It was a vital element in the city's economy, with its international airport being South Island's tourist gateway, a position earned and maintained with vigorous place promotion campaigns (Schöllmann et al. 2000). The growth in numbers of passengers coming through the airport peaked at about 6 million a year by 2010, the start of the earthquake sequence, but due to the loss of essential facilities, such as downtown hotels, and lack of access within the cordon, did not recover this level until 2016. Meanwhile, landings at Queenstown airport, which provides trans-Tasman flights with Australia, grew more quickly, although Christchurch recovered to seven million domestic and international arrivals in 2019, only to halve within two years due to Covid-19 (CIAL 2021; Queenstown Airport Company 2021). Some of the blueprint's anchor projects, notably the convention centre and multi-use arena, have to be seen in this context as vital elements of the city's place promotion strategies (Chapters 4 and 6).

The roll-out of these projects has been far from straightforward and exemplifies the difficulty of meeting objectives in complex administrative and political situations (Pressman and Wildavsky 1984). Lindblom (1959, 1979; also Johnson and Olshansky 2017) argues that in such situations, the best that can be achieved is a disjointed incrementalism, characterised as 'muddling through', and that it is therefore unsurprising that the best laid plans have not been delivered on time. In another way, the purpose of these central city anchor projects in generating and restoring consumption flows illustrates the principle of path dependency: the expectation that as recovery proceeds, consumers and tourists would return and progressively increase in number. As the airport statistics illustrate, this process has been slow, and as elsewhere slowed still further by the Covid pandemic (Škare et al. 2021). More disruption is likely with the increasingly urgent need to engage with climate change goals. This raises significant questions about the wisdom of depending on a path that locates major consumption facilities in the central city, requiring significant travel for consumers, although there are equally serious issues raised by provision of car-dependent facilities in the suburbs.

Retailing, hospitality and experiments in business

One of the attributes of pre-earthquake Christchurch was the long-term dispersal of retailing activities to the suburbs (Thull and Mersch 2005). This reflected the preponderance of suburban malls which in 2006 amounted to 0.82 square metres per person, larger than Auckland, at 0.63, and far in excess of Wellington, at 0.52 (Blundell 2006). The first such mall, and still the largest, was established in Riccarton in the 1960s; it now has the highest turnover of any in the country (Gibson 2019). The city had then, as now, three other big suburban malls, plus a host of smaller covered malls in specific districts, in line with its autodependent geography and relatively permissive District Plan (Chapter 5). While the larger malls are anchored by supermarkets, they are more than places to shop with an array of cafés, bars, restaurants and fast-food outlets and sometimes multiplexes. In the last decade, a growing number of big box developments have also been built, both in strips (for example, along the arterials Moorhouse Avenue and Blenheim Road) and at specific sites (as in Langdons Road, Papanui and the Shirley Homebase). The common feature that these facilities share is free car parking, along with those centres that have been master-planned within new greenfield subdivisions, such as Wigram and satellite towns, for example, Rolleston (see Chapter 5).

Post-earthquake central city shopping is now concentrated in the retail precinct of large-scale developments of 7500 square metres (Chapter 6), the product of the amalgamation of pre-earthquake land holdings, as required by the blueprint (Figure 9.1). These are integrated office, retail and car parking projects, although unlike at suburban malls, parking is a charge to users. They are mostly new builds although some, such as the Carter Group's The Crossing, preserve pre-earthquake facades. They include internal laneways and sometimes small squares, giving the illusion of an extending public realm to match the aspirations expressed by the public – as described in Chapter 4 – in Share an Idea in 2011. Combined with other infrastructure developments, such as re-made streets and the river promenade, this has created a varied and (significantly for a windy city) sheltered downtown environment for a more specialised range of consumption experiences than the serial replication characteristic of most suburban locations. Facilities in this rebuilt city centre are designed to capture the spending of domestic and international tourists, as well as office workers, and the discretionary dollars of those interested in specific hospitality experiences by day and night.

The central city's landcapes of consumption are therefore dominated by a range of big brand retailers, especially in fashion, which can afford the significant rents charged in expensive, post-earthquake buildings (Chapter 6).

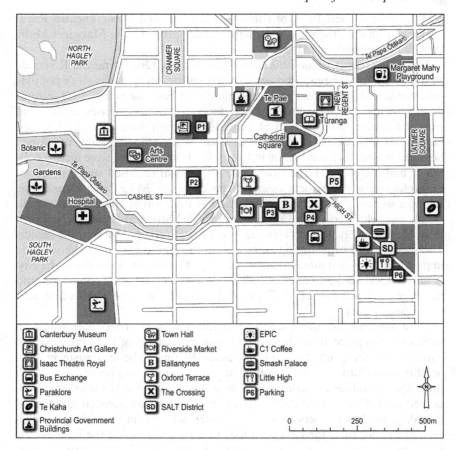

Figure 9.1 City centre consumption sites.

Smaller footprints are occupied by some more specialised shops, but also by a considerable number of food outlets, café-bars and restaurants. These are situated along the river, in the laneways and squares of larger developments, as well as within facilities such as hotels and Tūranga, the new central city library. There are however some specific locations that are the product of particular post-earthquake histories. For example, New Regent Street is a surviving streetscape built in the 1930s in Spanish Mission Revival style. Most of its owners and tenants clubbed together to restore their buildings, which have Category 1 historic place status. The facades were strengthened with structural steel frames, and most outlets in the street re-opened during 2013 (Naylor Love 2021), making this the first permanent retail and hospitality focus in the central city after the earthquakes. It sits well outside the retail precinct, to the north of Cathedral Square, somewhat in defiance of the cool logic of the blueprint. But its cafés, bars, restaurants and popular ice cream parlour have gained from proximity to the restored Isaac Theatre Royal (2014) and one of the most successful anchor projects, the Margaret Mahy Family Playground (2015).

Two more recent venues, the Little High Eatery (2017) and the Riverside Market (2019), have rehoused some of the businesses from the city's popular post-earthquake container mall (Chapter 6), embodying much of its spirit in developments that have brought new food experiences, senses of fun and experimentation to the central city. Little High houses eight licenced family-run businesses on the ground floor of an office building developed behind a grand, restored facade in Lower High Street. They were mostly container mall neighbours and now share a convivial enclosed space styled on examples overseas. The developers, Richard Peebles, Mike Percasky and Kris Inglis, also own the site that housed the last iteration of the container mall, which was relocated several times along Cashel Street as the new retail projects got underway. It is on this site that their Riverside Market now stands. The market, and surrounding lanes, is a more ambitious project, constructed in part from recycled materials from destroyed buildings elsewhere in the city. It has 30 food stalls, 40 market stalls and upstairs bars and restaurants, and has been such a drawcard that it hosted two million visitors in its first year (Hayward 2020a).

Developments like these seek to provide consumer experiences based on values such as place attributes, product quality, support of local business and a low carbon footprint. Riverside, for example, aims to be zero waste, in partnership with Cultivate Christchurch, an urban farm a kilometre away, which provides vegetables for sale as well as a destination for the market's composted waste (McDonald 2019c). The market's stalls are licenced rather than leased, giving low barriers to entry for traders who wish to experiment with a central city location for new products, thereby overcoming the issue of the high rents charged for more permanent sites elsewhere in the retail precinct. As such, the market is both an experiment in business practices and in place-making.

Another example of such an experiment is a collective initiative to rebrand the blueprint's Innovation Precinct as the SALT district. This initiative brought together some of Little High's (and the Riverside Market's) developers, with Ōtākaro Limited, the Crown company designated in 2016 with the completion of the anchor projects and ChristchurchNZ, the city council's development and promotions agency. SALT denotes the area around St Asaph, Lichfield and Tuam Streets, but also means 'South Alternative', and 'salt of the earth' (Figure 9.1). It won an international award as a large-scale place project, being described as 'an exemplar in community co-creation' (Place Leaders 2019). SALT is intended to denote something lively and gritty, an area of employment and entertainment that has been partly rebuilt around surviving structures in Lower High Street, and partly reconstructed as a venue for a range of businesses. These include the EPIC innovation hub, Little High, C1 Espresso and Smash Palace, an open-air transitional bar. To quote one of SALT's instigators, the central city 'can't compete with malls for retail or from the comfort point of view, but it can with *experiences*, so people go "wow, awesome"' (Percasky 2020).

The recognition given to the SALT project has led to the establishment of an unusual trust, Action Reaction Central Christchurch, that aims to encourage those working in different inner districts to highlight what makes their neighbourhood unique and use this to attract more consumers into the central city (Hayward 2020b). 'Uniqueness' in this context can mean such things as facilities, atmosphere, street art and events. In part, the formation of the trust is borne of frustration from the lack of coordination between public (especially city council events and ideas) and business initiatives. It aims to resolve what Marcus Westbury, who championed place-making strategies in the revival of the dying centre of Newcastle, New South Wales, describes as 'how places enable or thwart people with initiative' (2015, 12). The trust also seeks to recapture on the streets something of the spontaneity of the transitional activities that, as described in Chapter 8, flourished in the immediate post-earthquake period. These won for the city a lot of international notice such as for street art, featured in *Oi You!* Lonely Planet's (2017) recognition of international street art cities, and the multi-award-winning container mall, recognised by the Property Council of New Zealand and the Canterbury and national branches of the New Zealand Institute of Architects (Figure 9.2).

Figure 9.2 The SALT district: a three-dimensional street art illusion.

The most distinctive inner-city districts are those established as dining and entertainment destinations, such as SALT, Riverside and the Oxford Terrace 'strip' of bars. The last-named opened in 2018 as a complex of buildings and lanes, each with a different design, as a project by local developer Antony Gough. He has succeeded in reproducing one of the most vibrant of the city's pre-earthquake nighttime landscapes. In this sense, post-earthquake place-making builds on a decade's worth of experience in which an earlier city council worked with developers to inject consumption activity into an increasingly obsolescent downtown core. Gough has rebuilt in the same place; other pre-earthquake developments, such as Lichfield Lanes (on the fringes of what is now the SALT district), were housed in old brick buildings, most of which were damaged on 22 February 2011, and with one exception have been demolished (McDonagh 2017). But they pioneered in the city the encouragement of young people's drinking habits as a means of downtown revitalisation. This has thrived in 'an almost carnivalesque atmosphere' (Johnston 2014, 199), with two very effective consumption outcomes. It has spread patterns of use from business and family users in daytime and evening well into the night, as well as attracting a much more youthful – and some-times problematic – set of consumers.

The role of anchor projects in revitalisation

The rationale for the provision of anchor projects in the blueprint, as described in Chapters 4, 5 and 6, was to hinder capital flight by providing material evidence of the state's commitment to rebuilding the city, and to ensure that private sector investment was thereby encouraged. In reality, most of the main consumption-orientated anchor projects have been so delayed (Figure 4.4) that private investment, as described above, and in Chapter 6, has led rather than followed. Often this has been by taking considerable business risks. When the city's nighttime economy was seisimically destroyed within seconds, much of the activity relocated to the suburbs, for example along Riccarton Road to the west, or was absorbed into informal partying in pri-vate homes. Antony Gough's strategy for recapturing this trade with the con-struction of the Oxford Terrace strip nearly foundered with problems in accessing sufficient capital several times between 2013 and 2018. Developers like Mike Percasky (Percasky and Shaw 2020) have been vocal in their con-cern that the failure to complete the multi-use arena undermines their own investments. This section assesses the anchor project strategy in this light, focusing on the convention centre, the arena, the metro sports centre and the children's playground.

The convention centre, known as Te Pae (which has various meanings, including 'gathering place') opened at the end of 2021 after considerable delays, due in part to problems accessing skilled labour during Covid-19 out-breaks. It sits within Te Papa Ōtākaro/Avon River precinct, on a large site that required closing and erasing a length of one city street. With 28,000 square metres of floor space, it was planned in the blueprint as a complementary facility to the 32,000-square-metre New Zealand International Convention Centre in Auckland (due in 2024), and a smaller facility in Queenstown (yet to be built). In the meantime, however, Wellington is also building Tākina, an 18,000-square-metre centre opposite Te Papa Tongarewa, the national museum (due in 2023). Its business case is based on protecting the capital's convention market share once the other centres are open, as well as enhancing its 'international competititiveness' (Wellington City Council 2018, 7–8). This profusion led one journalist to describe a coming 'battle of the convention centres', each planned in pre-Covid times and when carbon management had not attained its present profile (Anthony 2021). They are all, however, city centre projects which can be unusual internationally.

Whether in this context the blueprint's promise of attracting 'new and exciting events' to post-earthquake Christchurch can be delivered remains to be seen (Central City Recovery Plan 2012, 67). Inevitably this will depend on competition at a range of scales, as recognised by Ōtākaro Ltd, the Crown company that built Te Pae, engaging ASM Global, a venue management company based in Los Angeles with a portfolio of more than 300 venues worldwide. The scale of ASM's ambition for Te Pae is shown in Figure 9.3. But ASM in turn competes with the city council's own business, Venues Ōtautahi, which manages the Christchurch Town Hall, on the opposite side of the river, and which used to be connected to the council's own exhibitions hall, lost in the earthquakes. The blueprint also questioned whether the Town Hall itself would be repaired, but as described later in this chapter, the

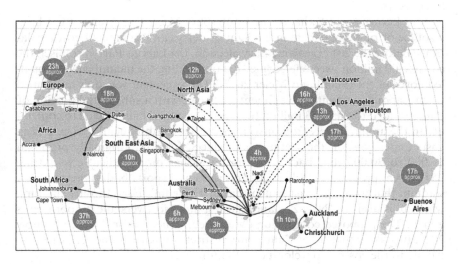

Figure 9.3 Te Pae: 'getting here'.

Source: Redrawn from www.tepae.co.nz

council decided otherwise. The council also declined to assume ownership of Te Pae in the Global Settlement Agreement with the Crown, signed in 2019, viewing it less as an asset than a long-term liability. In contrast, Woods Bagot (2021), the lead designer, asserts that it 'helps establish a strong cultural narrative throughout the central city', at best a speculative claim as it replaces in one outsize structure what were two lively pre-earthquake city blocks.

The earthquake sequence was well underway by the start of the year that Australia and New Zealand were to host the 2011 Rugby World Cup. The loss of the 38,500-seat rugby football and cricket AMI Stadium (located on Lancaster Park) to the earthquakes was therefore doubly felt by sports fans in Christchurch and Canterbury. The main stands were badly damaged, and the pitch destroyed by liquefaction. The city council redeveloped the Hagley Cricket Oval relatively inexpensively, as a traditional-style ground with pavilions and grass banks. It was finished in spring 2014, and used for the opening game of the Cricket World Cup, hosted by Australia and New Zealand, in February 2015. It has largely met the requirements of the cricketers (Longley 2015), although the later installation of night-time lights was more controversial. The Christchurch-based Crusaders Super Rugby franchise continues to operate in a small temporary open-air facility with a seating capacity of 18,000 spectators, inadequate from its point of view, given the quality of stadia in Dunedin, Wellington and Auckland. To meet their requirements, and those of concert-goers, the blueprint planners proposed the construction of a large 35,000-seat covered multi-purpose sports and entertainment venue in central Christchurch, much closer in form to other cities' facilities than Lancaster Park had been. It was to host rugby union, rugby league and football up to an international level, and also allow for entertainment events such as concerts.

Now known as Te Kaha, or the Canterbury Multi-Use Arena, it has been a source of ongoing debate in the light of delays and significant cost escalations. Before the 2017 parliamentary election, the Labour Party promised the city a $300 million sweetener, somewhat outbidding the incumbent National-led government (Small and Meier 2017). The sweetener was duly confirmed when Labour came into government that year, with a substantial portion of it to be added to the $253 million that the city council had decided to contribute to the arena, largely from insurance receipts. Subsequently, to help keep costs under control, the seating was reduced from the originally proposed 35,000 to 30,000, but back up from an earlier more radical cost-cutting 25,000-seat proposal (Clark 2021). Investors and sports bodies pressed for a stadium that would at least match that of Dunedin's Forsyth Barr stadium, which bills itself as 'New Zealand's largest and most versatile indoor events arena'. The proposed completion date for construction for the arena is now 2026; it is likely that other Canterbury district councils will be asked to contribute to the costs of development (Walton 2021d).

Surprisingly, given the capital outlay and likely ongoing costs of the facility to Christchurch and Canterbury ratepayers – local property taxes are predicted to increase significantly – there was for long little open opposition to the proposal. But it began to emerge with further cost escalations in mid-2022 and may yet match that experienced from 2009 in Dunedin where the scars of the battle over its stadium were still obvious a decade after it was proposed and then built (Loughrey 2019). What has been largely missing from the discussion in Christchurch is that while a significant literature identifies the great enjoyment engendered by arenas among their users, and the players in, and owners of, professional sports teams, it also points to a very high and perhaps unsustainable level of public subsidy and cost in support of considerable private gain (Swindell and Rosentraub 1998). But the case both for and against is always hard to assess in such situations: in the last decade, it has been Dunedin, the South Island's only other significant city, that has attracted All Blacks rugby games and big international acts, such as Ed Sheeran, Elton John and Queen.

The blueprint's metro sports facility, now known as the Parakiore Christchurch Recreation and Sport Centre, is another of the large anchor projects plagued by completion delays (McCrone 2015b). Designed to complement a range of new post-earthquake suburban physical recreational facilities (described in Chapter 10), it will replace the liquefaction-damaged eastern suburbs QEII Park as the city's main swimming complex, and provide Christchurch's netballers with a long-needed indoor home. The centre was supposed to be completed early in 2016. Six years later, it was still under construction on a four-hectare central city site vacated by a large brewery after the earthquakes. Once finished it will be the largest aquatic and indoor recreation and leisure venue in New Zealand and be accessible to people of all ages, abilities and skill levels. In its more than 30,000m² space, the Centre will contain a 50 metre, ten-lane competition swimming pool and a separate diving pool, a large aquatic leisure area, five hydroslides, fitness spaces and nine indoor courts for sports such as netball, volleyball and basketball. It will cater for recreational, educational and high-performance sports communities, although tensions over priority between these groups quickly emerged.

In 2015, journalist John McCrone observed that 'just like the convention centre, the East Frame, and some of the other anchor projects…behind the scenes it [the Recreation and Sports Centre] has turned into something of a monster' (McCrone 2015b). Progress has been hampered by a range of complex administrative difficulties. Cost-overruns have occurred, partly because, as sports administrators see it, early plans bordered on the grandiose. The delays have caused a great deal of frustration among sports groups in need of the promised facilities. In the interim, both high-performance sports groups and recreational participants have sought and relied on temporary facilities – some in other provinces with, for example, swimmers travelling 360 kilometres south to Dunedin (McCrone 2015). In light of the cost-over-runs, amounting to $NZ75 million by November 2017, Dr Megan Woods, the new Labour-led government's Minister for Greater Christchurch

Regeneration, announced fresh administrative, design and contracting arrangements (Woods 2017). Inevitably, this took time, so the first sod on the site was not turned until June 2019. The intervention of the Covid-19 pandemic and subsequent labour shortages further delayed the opening.

In contrast, the award-winning Tākaro ā Poi/Margaret Mahy Family Playground was the first of the major anchor projects to be completed, a year after the Hagley Cricket Oval, in late 2015 (IPWEA 2019). It is part of the Te Papa Ōtākaro/Avon River precinct and named for Christchurch's internationally recognised children's author, and also honours another remarkable Canterbury writer and community activist, Elsie Locke, whose city council memorial park was destroyed in the earthquakes. The playground incorporates landscape themes developed in conjunction with Ngāi Tahu and was co-named by them. It is large, covering a 2.5-hectare site, and the biggest project in New Zealand created primarily for play (Hamilton 2016, Wade 2018). Somewhat controversially, the playground cost more than $NZ40 million to construct (Smith 2017). It is however hugely popular with children, their families and young adults, the latter using it in the early evenings and after dark (Figure 9.4).

The playground had its origins in the city council's Share an Idea project and a close engagement with Christchurch's children. In 2013, the Central City Development Unit initiated a competition inviting children to design 'the world's best playground'. More than 6000 children participated sending in approximately 300 entries (Stewart 2013). The task of operationalising the entry from Year 6 at Selwyn House School – which suggested the

Figure 9.4 Children and adults in the Margaret Mahy Family Playground.
Credit: ChristchurchNZ.

Margaret Mahy naming – was given to a consortium comprising a play safety specialist, a playground supplier company, materials engineers and a team of designers. Their objective was to create a design that catered for a variety of age groups marked by different levels of challenge using multiple play media. And as one journalist wrote after it opened, 'If you're in the mood for a physical challenge, a race, climbing, digging, squirting, bouncing, sliding, spinning or skimming, this is the place to lose time' (Wade 2018). It has won a number of awards, including top honours as a public works project by the Institute of Public Works Engineering Australasia in 2017, and a New Zealand Institute of Landscape Architects Award of Excellence in the same year. For some years, it stood apart in meeting the anchor project aspiration to attract people to the central city and contribute to business growth (Hamilton 2016; Smith 2017). Wotif, the online accommodation platform, advertises a range of hotels close to the Armagh Street site that can accommodate visitors to the playground (Wotif 2021).

Cultural heritage and restorations

In contrast to new anchor projects designed to attract consumers to the city centre, many of its cultural heritage buildings, 'in an architecturally serious city in which the bar was set early, and high, by … High Gothic Victorian revivalism' (Walsh 2020, 117) were destroyed in the earthquakes or written off by the blueprint. Some major facilities outside the blueprint area nonetheless fared relatively well. The city's Botanic Gardens were largely unscathed, as was the Gothic Canterbury Museum which had been seismically strengthened before the earthquakes, although it had long yearned to modernise its warren of buildings to better comply with contemporary museum practice. The Arts Centre, the largely Victorian campus vacated by the University of Canterbury in the 1970s, was less fortunate, with every building damaged. But insurance proceeds have underwritten a massive restoration project. The nearby Art Gallery (2003) functioned as the headquarters of Civil Defence during the major quakes, but turned out to need major repairs before re-opening in 2015. The fate of the Christchurch Town Hall was more contentious.

The contest over its future is an example of the performative power of the map proving ultimately impotent. A modernist composition of interconnected buildings on the river banks, the Town Hall includes a concert hall widely renowned for the quality of its acoustic and audience experience. Leonard Bernstein, conducting the New York Philharmonic there in 1974, two years after its opening, declared that he was 'very envious … I wish we had something like it in New York' (Lochhead 2019, 125–7). More recently, the *Guardian* newspaper included it amongst '10 of the world's best concert halls' (Cox 2015). The complex was badly affected by lateral spread on 22 February 2011 and was omitted from the blueprint, in favour of facilities in a new Performing Arts precinct. The Minister for Earthquake Recovery at the time, National's Gerry Brownlee, was an implacable opponent of restoration, ignoring affection for 'Christchurch's public living room'. In this case, however, the map did not win out; the Town Hall was re-opened in 2019, not least because a business case prepared by international accounting firm Deloitte for the owner, the city council, estimated restoration would cost far less than a new facility (Law 2015).

How the Town Hall will fare operationally, with the opening of competing facilities in Te Pae and the Performing Arts precinct, remains to be seen. The precinct is effectively anchored by the Isaac Theatre Royal and New Regent Street, both of which, as indicated earlier in this chapter, were restored fairly quickly. The precinct was intended to provide performance facilities for the Christchurch Symphony Orchestra (but now, as before, these are based in the Town Hall), the Christchurch Music Centre (previously in the destroyed Catholic Cathedral grounds), and the Court Theatre (previously in the Arts Centre) (Central City Recovery Plan 2012, 77). A new facility, The Piano, was opened for the Music Centre in the precinct in 2016, funded by insurance payments with contributions from the council and Crown (McDonald 2016). Negotiations about the future of the Court Theatre have been more protracted. It is the country's largest professional theatre company, and decamped to a converted industrial warehouse in the suburb of Addington in late 2011. This facility is greatly enjoyed by theatregoers because of its spaciousness, acoustics and on-site car parking. Following offers of council and Crown support, it is expected that a new theatre will open in the precinct in 2024, although this still left some space unallocated.

The provision of new facilities, sometimes but not always aligned with the restoration of the old, raises questions over the affordability let alone use of repairable, older cultural heritage buildings. Repair was not usually considered an option in the blueprint, and in many cases was not viable. Some smaller buildings have been restored by private developers who, in terms of the argument advanced in Chapter 6, have been attracted by the business opportunities that can be generated from enhancing some sense of place in the urban landscape. Companies such as Box 112 and the Stockman Group have been active in this respect. Box 112 has re-purposed a number of historic buildings in the river precinct, including the majestic former Public Trust Office, as well as former industrial and workshops spaces for small-scale retail, hospitality and personal care businesses in the district now dubbed 'South Town'. Stockman has restored a number of old buildings in Lower High Street, as has Richard Peebles. These restorations have been more practical for both developers and tenants than larger-scale projects which were often questionable business propositions even before the earthquakes.

The result of the priorities set within the blueprint has therefore been that over a decade later, some significant restorations remain incomplete, or even un-started. This is sometimes due to the complexity of repair, but often down to a shortage of available funds. The blueprint required heavy commitments from both the city council and the Crown, and one-off commitments to iconic special cases, such as the Anglican cathedral restoration, discussed in Chapter 4, have left little enthusiasm for further demands on the public purse. The cathedral, which is the responsibility of the Church Property Trustees, still has a fundraising target of over $50 million, without taking account of rapid rises in building costs since the start of the pandemic in 2020. The Canterbury Museum has a $70 million shortfall for its modernisation plans and seeks to bridge this in part by seeking support from other Canterbury district councils, as is the case for the new multi-use arena discussed above. The Arts Centre, overseen by its own Trust Board, has insufficient funds to complete the restoration of a third or so of its buildings (Law 2021). These buildings are each ranked as New Zealand Historic Places Trust Category 1.

The problem that such facilities pose is exemplified by the Canterbury Provincial Government Buildings, before 2010 New Zealand's only surviving complex from the provincial period of national government (1853–76), and also Category 1. It sits as a ruin on the opposite side of the river to Te Pae. The timber parts of this Gothic complex, built in the late 1850s, survive more or less intact, but the later stone towers, and a magnificently flamboyant stone council chamber completed in 1865, crumbled in the 22 February earthquake in 2011. It had long been hard to find effective uses for much of the complex. Although the ruins were quickly sealed against the weather, and the stones carefully stored, it is not clear when or if restoration can begin. In 2018, it was estimated that this would cost over $200 million, but that the city council, as owner, could commit only $20 million of this by 2028, with the balance not until between 2032 and 2040 (Law 2021). A proposed targeted property rate is unlikely to make much difference. The only conclusion that can be drawn with any certainty is that recovery after a great disaster takes a very long time.

In contrast to these concerns for the loss and regeneration of European New Zealand-built heritage has been the work of the Matapopore Charitable Trust and its mission to highlight cultural diversity in the Christchurch urban region. A collaboration between local Ngāi Tahu hapū (sub-tribe) Ngāi Tūāhuriri and originally CERA, and now the region's local authorities, the trust has been responsible for representing Māori values, narratives and aspirations in the heritage recovery process. The trust has made its presence felt strongly through the widespread use of Māori designs and te reo (Māori language) names for new buildings in multiple urban sites during the recovery. These have included motifs and poems etched onto buildings and pathways, new public artworks, and gardens including indigenous and endemic plants, and appear alongside and sometimes interwoven with established landscapes, all of which contribute a new element to consumption in the city that is already appearing in Christchurch and Canterbury's tourism promotion (Harvie 2016; McDonald 2018c).

Conclusion

One of the key threads of this chapter has been the importance of consumption, leisure and the provision of landscapes that encourage consumption practices and experiences in post-earthquake recovery. The focus has been on those forms of consumption at the centre of the rebuild: the ever-expanding range of commercially provided 'objects', that includes physical things, services, practices, cultural performances, lifestyles and places, and the public recreation facilities and spaces they complement. It has paid less attention to outdoor recreation and engagement with the more-than-human world, which are discussed in Chapters 11 and 12. Instead, we conclude here with a brief reflection on two themes: first, that work to attract flows of consumption is highly competitive, and second, that such flows often depend on and drive a further cycle of consumption, that of carbon.

Competition to attract consumers and consumption practices operates at a range of spatial scales. An important part of the state's support for the rebuild of the Christchurch urban region and its focus on consumption has been to maintain the economic role of the South Island's largest city and its tourist gateway. The central city blueprint was designed to hold and attract investment and encourage people to visit the central city to engage in a wide range of leisure activities. The role of the anchor projects within this was to be that of leading private sector investment, even if generally it has worked in practice the other way round. While local residents have been attracted to the regenerating inner city, Covid-19 has complicated the assumption that it would also recover steadily as a flourishing site for national and international tourism.

In taking this approach, the blueprint has laid the basis for future competition between Christchurch and other New Zealand urban centres, for instance in the provision of convention centres and multi-use arenas. It has also underwritten competition between the city centre and the suburbs in retailing and hospitality, and to an extent through the centralisation of sporting and play facilities. This is not a clear-cut picture, particularly in retailing, and as Chapter 10 illustrates, there have been attempts, in the provision of new schools and swimming pools, to counter the lack of attention given to investment in the eastern suburbs of the city. But competition for resources has also emerged between initiatives, shown most clearly in the lack of funding left, after the choices made to spend so much on the anchor projects, for restoring key elements of cultural heritage.

The second theme that underlies an understanding of the city's new landscapes of consumption is that of carbon. Centralising facilities in the central city requires people to travel over an increasingly wide area, as a pre-occupation with the provision of car parking illustrates, despite some genuflection towards rebuilding 'a green city' in both the council plan and the Crown's later blueprint. This did produce early completion of one anchor project, the Bus Xchange, in 2015, but despite this, public transport patronage has yet to recover to pre-earthquake levels. And if the city and suburbs now have an emerging network of cycleways, encouraged by government funding available nationwide over the last decade, this has also yet to make much difference to travel patterns in the round. Indeed, one of the contradictions of the rebuild, as Chapter 5 so clearly shows, has been the facilitation of suburban and sub-regional greenfield expansion, supported by investment in Roads of National Significance ringing the city (Pawson 2022). In many respects, the rebuilt city typifies a carbon-intensive approach.

10 The eastern suburbs

Introduction

In the years following 2010 and 2011, there was some disquiet regarding the pace of recovery in the eastern suburbs of Christchurch. Attention was drawn to the extent of unrepaired property damage, to the apparent indifference of central government to community concerns and to ongoing problems with water and sewage infrastructure. Unfavourable comparisons were made with the more affluent western areas of the city, where building and infrastructure repairs were further advanced, and with the central city, where the government had invested heavily in a series of anchor projects and thematic precincts, and new landscapes of consumption were appearing. The notion of the 'forgotten east' began to circulate (Conway 2011; Lynch 2019; Smith 2018), encapsulating the view that eastern areas of the city were somehow being overlooked or under-prioritised in the post-quake recovery.

Responding to this concern, this chapter considers whether the eastern suburbs were in some respects neglected during the recovery process. It recognises that these suburbs were extensively affected by the earthquakes, with many areas suffering significant land damage and liquefaction. If the pace of recovery was slower here than elsewhere in the city, perhaps this simply reflected the large volume of work needed to repair the damage. Some of the city's most socially and economically disadvantaged communities are in the east, and such groups are known to fare worse during disaster recovery, whether on the basis of socio-economic status or race/ethnicity (Cutter et al. 2006; Enarson and Morrow 1988; Gaillard et al. 2017; Kahn 2005; Wisner et al. 2004). So, this too may have been a factor, at least in terms of the psychosocial dimensions of recovery. In the discussion here, however, the focus is on the actions of government, insurance companies and community organisations involved in the recovery of the eastern suburbs. To what extent were these actors responsive to local concerns? Were their efforts appropriately attuned to local needs, or were they in some way neglectful? Through a consideration of available evidence, this chapter seeks to address these questions.

DOI: 10.4324/9780429275562-13

Disruption and recovery in the eastern suburbs

Although there is some variation in local understanding, the eastern suburbs in Christchurch are generally understood to include Richmond, Avonside, Dallington, Wainoni, Burwood, Linwood, Phillipstown, Bromley, Aranui and New Brighton (Figure 10.1). Many of the communities in these areas are of lower socio-economic status, with higher proportions of low-income households and below-average levels of post-secondary education (Table 10.1). There is also above-average 'blue collar' employment (eg. labouring and manufacturing). Families with multigenerational connections to their neighbourhoods and thus significant place attachments are common. The largest ethnic group by proportion is New Zealand European, but a number of areas have significant proportions of Māori households, particularly Aranui and Wainoni. Given their proximity to the Avon River and development on drained wetlands, many eastern suburbs suffered extensive liquefaction and land damage in the September 2010 and February 2011 earthquakes. Residential water and sewage systems were also disrupted, with some neighbourhoods – in areas such as Bexley and New Brighton – having to rely upon portable toilets for weeks and, in some cases, months (Potangaroa et al. 2011).

There is a broad range of possible outcomes for places affected by disasters, as Wilson (2013, 2014) has observed. After the initial disruption, some manage to recover something approximating their previous levels of social and economic activity. This may reflect their internal resilience as well as the

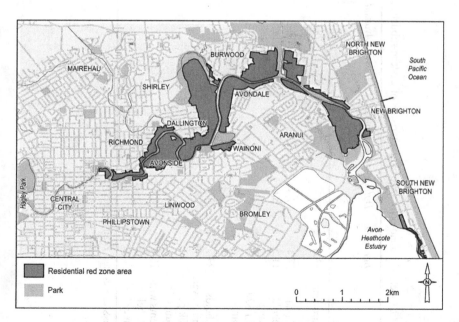

Figure 10.1 The eastern suburbs and the residential red zone.

Source: https://opendata.canterburymaps.govt.nz/

Table 10.1 Selected socio-economic characteristics of eastern suburbs

	Burwood	Dallington	Wainoni	Aranui	Avonside	Linwood	Phillipstown	Bromley	New Brighton	Christchurch average	NZ average
NZ European (%)	83.6	82.7	77.1	62.1	78.2	64.8	64.8	75.2	82.3	80.1	70.7
Māori (%)	7.4	7.4	12.4	17.9	12.5	9.2	13.8	13.3	9.4	6.7	11.7
Households earning under $20k per annum (%)	11.5	8.6	11.6	14.8	17.2	14.9	13.2	9.8	12.3	9.0	9.4
Households earning over $100k per annum (%)	23.7	13.9	10.1	8.0	12.6	8.0	7.1	10.6	12.9	23.8	23.4
Residents with a Bachelor degree (%)	10.4	8.6	3.8	3.1	7.7	9.1	7.7	4.2	9.2	12.6	12.1

Source: Statistics New Zealand (2013).

receipt of external support. Others places stagger onwards, continuing to function but in an appreciably diminished manner. There may be reduced levels of economic activity, increased material hardship, damaged infrastructure and residents grappling with personal and collective trauma (Fullilove 2016; Herman 1998). In places very starkly affected by disasters, people may be forced to leave, for reasons that include preservation of life, compromised housing, lack of local employment opportunities and grief and emotional loss that become too difficult to bear (Fullilove 2016; Pain 2019). For those who stay, the provision of meaningful support and access to relevant resources is a key determinant of their post-disaster experience. This is one reason why it matters a great deal whether the needs and concerns of places are recognised by key actors in the recovery process. If in the aftermath of a disaster those needs are somehow forgotten, disregarded or overlooked, then recovery may be compromised. For places as well for people, being noticed and attended to has substantive consequences.

Guided by these perspectives, the following sections examine four aspects of recovery in the eastern suburbs: the availability of affordable housing, the creation of the residential red zone, the rebuilding of sport and recreation facilities and the reorganisation of public education. As the repair of subsurface infrastructure was a city-wide issue, and not something specific to the eastern suburbs, it is not discussed here. For each of the four aspects, consideration is given to the ensemble of organisational actors involved – including central and local government agencies, insurance companies and community groups – and to whether they recognised and responded to residents' needs. By considering the extent to which the needs of the eastern communities were attended to, the notion of the 'forgotten east' is assessed.

Availability of affordable housing

Prior to the earthquakes, there was a visible concentration of low-cost housing east of the central city, extending into suburbs such as Avonside and Linwood. Many of these units were privately owned but some belonged to the city council, reflecting its longstanding commitment to providing housing for low-income residents. A proportion was also owned by Housing New Zealand, the central government housing agency which was restructured to become Kāinga Ora – Homes and Communities in October 2019. Irrespective of their ownership, these properties provided affordable housing for a range of people who had limited incomes, including individual men and women. Some were employed in low-paying jobs and others were long-term beneficiaries (eg. as a result of chronic illness, injury or disability). Around 600 people are estimated to have lived in the privately owned units, which included bedsits and small flats for single-person households (Kane and Smith 2013).

Much of the low-cost housing in the inner city east was badly damaged during the quakes, with an estimated 250 low-cost private rental beds rendered unsuitable for inhabitation. In this situation, many tenants were forcibly dispersed. Although there is no comprehensive research available on the housing biographies of those displaced, the Christchurch rental housing market at this time was undoubtedly challenging. Earthquake-related property damage had reduced the supply of rental accommodation, driving up market rents and the significant influx of tradespeople involved in reconstruction added further demand. It was not until the greenfield housing developments in the southwest of the city were completed some years later that the supply of rental housing improved and rental prices began to decrease (Chapter 5). Prior to this, the council had noted that 'adequate and affordable housing is a pre-existing challenge that is now a major recovery issue for the city' (Christchurch City Council 2013, 17). The number of people in insecure housing – a category that includes those struggling to meet rent payments, as well as those staying on a temporary basis with friends and family – was estimated to have risen from 3750 before the quakes to between 5510 and 7405 (an increase of between 46 per cent and 98 per cent) (Ministry of Business, Innovation and Employment 2013).

The landscape of affordable and low-cost housing in the eastern suburbs evolved in several key respects after the quakes. In the private rental sector, there was no coordinated or extensive redevelopment of the damaged and demolished units in the inner city east. Housing New Zealand paid for displaced social housing tenants to stay in motels across the city on a temporary basis. Although this arrangement provided them with a degree of stability, the motels were often located well outside of the eastern suburbs, which disrupted the connections people had to their original neighbourhoods. For families, the transition to motel living also generally meant a significant reduction in living space and the loss of an external garden. Over time, however, Housing New Zealand did oversee the construction of replacement social housing. New developments were completed in Richmond, an area with existing services and reasonable public transport connections, and in Aranui, which now has the highest number of Kāinga Ora homes of any Christchurch suburb (Walton 2021a).

By mid-2021, there were around 8900 social housing units across Christchurch. Around 6200 (70 per cent) were owned by Kāinga Ora, with concentrations in suburbs east, northwest and west of the city centre. Another 1944 (22 per cent) were owned by the city council and typically managed by the Ōtautahi Community Housing Trust; these units were also distributed across the city, with pockets in the south and inner city east, including Linwood, but few in other eastern suburbs. The trust owned another 500 units itself, and the remainder were provided by faith-based organisations such as the Salvation Army and Methodist Mission. In geographical terms, the city council's social housing provision in the eastern suburbs is slightly reduced compared to before the earthquakes, whereas the involvement of Kāinga Ora has expanded. The number of private rental units in the inner city east has not returned to previous levels. This reflects the responses of multiple private property owners to quake-related damage and market conditions, however, rather than a specific policy setting of central or local government.

The residential red zone

After a programme of geotechnical investigations, the Canterbury Earthquake Recovery Authority (CERA) announced in June 2011 that all residential land in Christchurch would be assigned to one of four colour-coded zones: green, orange, white or red, as described in Chapter 3. These zones reflected the government's assessment of whether a parcel of land was suitable for ongoing residential inhabitation, with particular regard to the risk of future damage from seismicity. In green-zoned land, the probability of this damage was considered low, so householders were expected to repair their homes as needed. In red-zoned land, however, the risk of further and future damage to land and property was judged to be unacceptably high. As a result, red-zoned homeowners were encouraged to leave their homes permanently, incentivised by a government offer of financial compensation, made initially in August 2011. But the offer itself was far from clear cut, as owners were given two options: sell their land and house to the government at the 2007/8 rating valuation (this being the most recent city-wide valuations available at the time), or sell only the land to the Crown and negotiate a price for the house with their insurance company (CERA 2011; MacDonald and Carlton 2016). In addition, the offer of financial compensation was only available to fully insured households, with the Minister for Earthquake Recovery insisting that uninsured homeowners should receive only 50 per cent of the government compensation offer. This reflected the fact that uninsured properties did not qualify for Earthquake Commission cover, so the Crown would gain no insurance compensation from that source.

The creation of the residential red zone (Figure 10.1), which encompassed over 5500 households, provoked a variety of responses. The 2011 market value of some red-zoned homes was significantly higher than the 2007 valuations used as a basis for the Crown offer of financial compensation, and the offer did not recognise any home improvements made in the meantime. For many homeowners, accepting the offer thus meant an appreciable financial loss (Matthews 2020). Some red-zoned residents were nevertheless relieved to receive the offer – especially when they had been grappling with their insurance companies for some months – as it enabled them to depart in a relatively clear-cut manner and move on. This, in fact, was the outcome the Crown hoped its offer would incentivise (CERA 2011). Others were angry that the financial compensation was below estimated 2011 market values. Still others complained about the erosion of private property rights and democratic process. These responses were by no means mutually exclusive, and some homeowners identified with them all. The collective sense of grievance prompted the creation of several community activist and protest groups, including Quake Outcasts, WeCAN, Empowered Christchurch, CanCERN (Canterbury Communities' Earthquake Recovery Network), and Eastern Vision, all of which advocated in different ways for the interests of red-zoned and eastern households.

Quake Outcasts, a group comprised of 68 of the uninsured red-zoned homeowners, and WeCAN took a case to the High Court to contest aspects of the red-zoning decision (see Chapter 7). The court found that the Minister for Earthquake Recovery's decision to discriminate against uninsured buildings had been unlawful, and also found that the red zoning itself was unlawful. This latter element of the decision was subsequently overturned by the Court of Appeal in 2017, but the judgement that the government's discrimination against uninsured red zone residents had been unlawful was upheld. Despite numerous submissions and public meetings organised by these community and activist groups, the National-led government's Minister for Earthquake Recovery at the time, Gerry Brownlee, had refused to admit any deficiencies in the red zoning process. Nor would he acknowledge the serious problems that many households within the red zone were experiencing when dealing with the Earthquake Commission regarding their insurance claims. Many residents reported apparent bureaucratic indifference to their concerns, while having to negotiate confusing and sometimes inconsistent communications from both the Commission and their private insurance companies (Miles 2016). Mike Coleman, the Anglican priest and Avonside resident who led WeCAN (Figure 7.3), described the situation as people 'battling a government who are totally shut down to affected homeowners' (Matthews 2020).

Within a year of the red-zoning announcement, most affected residents had nevertheless reached a settlement with the government for their homes and land. Some relocated within Christchurch, buying homes in the new subdivisions in the southwest of the city if their finances allowed, or to the Selwyn and Waimakariri districts outside the city (Campbell 2014). Others moved elsewhere in New Zealand or to Australia, sometimes prompted by a desire to escape the ongoing difficulties and stress of the post-quake situation (Adams-Hutcheson 2015). As they left, the population of the Ōtākaro-Avon River Corridor statistical area, which incorporates the majority of the red zone, fell dramatically. In 2006, it had a population of 10,386, but in 2013 this was just 1263, and by 2018 had further decreased to 99. This outmigration disrupted longstanding neighbourhood relationships, fashioned through everyday activities such as walking children to school, participating in neighbourhood watch groups, attending occasional street parties, sharing home-grown vegetables, lending and borrowing tools, crowding around a television to watch sporting matches of national importance, and so on.

In most cases, vacated red-zoned houses were demolished, but a few people made arrangements for their purchase – from the Crown, which had taken ownership – and then relocation to sites elsewhere. Either way, the eventual effect was to create large swathes of open greenspace (Figure 10.2). Initially, the property boundaries remained quite visible, particularly where these had been marked by trees or shrubs. Over time, the removal of street signs and other orientating features in the neighbourhoods made it harder to identify individual sections, and the more obvious visual markers of where the homes had once been faded. More-than-human actants became increasingly active, as the next chapter illustrates. After the majority of houses in red-zoned areas were cleared, a number of the affected communities began to hold occasional commemorative events. These gatherings, organised from 2012 onwards by

Figure 10.2 Vacated land in the residential red zone (aerial and street views).

Credit: Stuff Limited (aerial view) and Michal Klajban, Wikimedia Commons (street view).

the Avon-Ōtākaro Network, provided an opportunity for previous and current residents to meet again and to continue the process of coming to terms with their forced displacement.

In 2019, the occasion was held in a park near Avonside, and included elements of memorialisation with an exhibition of photography by a local artist, who had worked in collaboration with the community (Veling 2021). It was opened by the Mayor, Lianne Dalziel, who had previously been a Member of Parliament for the Christchurch East constituency. There were contributions from former students of several affected schools and a perspective from a local Avonside resident. After the speeches, a community barbecue was held, enabling those present to interact more informally. Such activities resonate with the kinds of 'memory work' that Till (2012) identifies as potentially reparative in 'wounded cities', particularly when places have been affected by overlapping forms of environmental damage and bureaucratic difficulties. The opportunities for community narration in Avonside also align with Herman's (1998) recognition that practices of remembering and retelling can help in the negotiation of collective trauma. The collective sharing of stories can assist individuals, households and communities to make some sense of what has happened, both to their places and their lives, and to craft narratives that afford points of stability, however provisional, from which to move forward.

Such memorialisation can be considered part of the broader creative response to the Canterbury earthquakes. In an effort to chronicle the lived experience of disruption, displacement and recovery, a number of novels, non-fiction accounts and collections of poetry have emerged (eg. Farrell 2012, 2015; Gordon et al. 2014; Gorman 2016). An example is *Leaving the Red Zone*, a collection of 148 poems by more than 85 individuals (Norcliffe and Preston 2016). At its launch, several local government officials from within and beyond Christchurch attended, including the mayor (Preston 2016). Their presence perhaps reflected a recognition of the value of creative expression in the ongoing process of coming to terms with the earthquakes. Although such expression is not neatly coincident with recovery, it can assist with the sense-making and reorientation that supports people to deal with loss and displacement (Herman 1998).

In 2020, some nine years after the announcement of the red zone, the Global Settlement Agreement between the Crown and the council, described in Chapter 4, sealed the transfer of the 602 hectares of red-zoned land to the city. This mirrored what had happened in the Kaiapoi red zones in 2016. The transfer was intended to proceed in an incremental fashion as Land Information New Zealand, the Crown holding agency, completed the amalgamation of land titles. Expressing a preference for local ownership, the Labour-led government's Minister for Greater Christchurch Regeneration, Dr Megan Woods, described the process as 'the next step in the transition to local leadership ... The land is now a beautiful park-like space that has the potential to become a true asset for our city and something that we as Cantabrians can all be proud of' (Christchurch City Council 2020). The intent to transfer the red-zoned land was widely regarded as an important step in returning the power that central government had taken from the city council in the immediate aftermath of the quakes. As such, it arguably represented the return of local land into more directly accountable and democratic arrangements.

Sport and recreation facilities

Alongside the loss and displacement associated with the residential red zone, a number of publicly owned sport and recreation facilities in the eastern suburbs were heavily damaged by the earthquakes. The majority of these were eventually repaired or rebuilt, although generally at a smaller scale (and, in some cases, key elements of the original facilities were moved out of the area entirely). When it occurred, this 'downsizing' and 'relocation' often reflected the westward shift in the city's population, as well as the financial constraints of local government in the aftermath of a major disaster. The most materially and symbolically significant of these facilities was the Queen Elizabeth II (QEII) swimming pool and athletics complex in Parklands. Built for the 1974 Commonwealth Games, this large facility incorporated the city's only indoor 50-metre pool, other recreational pools and several hydroslides. It was an important venue for both the swimming and athletics communities, with a high-quality outdoor track and field facilities that supported regional and national competitions.

The QEII facility sustained serious damage in the February 2011 earthquake, and in April 2012 it was served with a demolition notice by CERA. A smaller and wholly indoor facility, the $38.6 million Taiora QEII Recreation and Sport Centre, was eventually built on the same site, opening in May 2018 (Figure 10.3). Its completion was described as marking the 'end of a long and hard-fought battle by the community to get a pool rebuilt at QEII Park, after the much-loved former complex was demolished' (Law 2018). Its reconstruction

Figure 10.3 The new Taiora QEII recreation and sport complex.
Credit: Simon Wilson.

had by no means been certain, particularly after central government decided to jointly fund, with the city council, the Parakiore anchor project facility in the city centre, as described in Chapter 9. The size, central location and Crown funding for Parakiore undermined the case for rebuilding QEII in the eastern suburbs. Could the council afford to be involved in building two large aquatic sport complexes? Mindful of such issues, a community group known as 'Keep QEII in the East' lobbied the council for several years, arguing that rebuilding a facility in the eastern suburbs was critical if the city wished to be receptive to the needs of its eastern residents.

When the Taiora QEII facility was completed in 2018, seven years after the most damaging earthquake, it was widely regarded as confirmation of the city council's responsiveness to the needs of the area. In the media coverage of the opening, the material benefits and symbolic significance of the development were recurrent themes. As one woman explained, the return of the facility to its former location had 'put the east back on the map. It's recognised the families of the east [and their] wellbeing and health and that community link which, at the end of the day, always stood staunch through all the quakes' (Ineson 2018). That said, the Taiora centre did not retain any of the outdoor athletics facilities of the original QEII. These were effectively transferred to the Ngā Puna Wai Sports Hub, a new $53.7 million multi-sports facility in the southwest of the city, located within easy reach of the extensive new suburban housing developments in Halswell, Lincoln and Rolleston (Chapter 5). This reflected the westward shift in the provision of city-wide public facilities, with the council unambiguously describing Ngā Puna Wai as the 'new sporting home for Canterbury athletics, hockey, tennis and rugby league', noting that it was intended to 'replace sporting facilities damaged in the earthquakes' (Christchurch City Council 2021b). Nonetheless, the Taiora centre can still be regarded as a firm local government commitment to the city's eastern suburbs.

The new centre has proved very popular, drawing over 800,000 visitors during its first year of operation to its aquatic, gym and group fitness activities. This included about 540,000 visitors to the pools alone. The design brief of the architects had been to 'create a sustainable and public building that restores and grows participation in sport in the east and the wider Christchurch community', and the initial signs are that the rebuilt Taiora will fulfil these aspirations (Warren and Mahoney 2021). The council also subsequently built Te Pou Toetoe, a new $22 million pool and activity complex in Linwood, which opened in October 2021. This facility also includes two tennis courts and a basketball court. Its development recognised the need to replace the earthquake damaged pool at Linwood Avenue Primary School. It also responded to the strong call from eight local school principals for a new public swimming facility (Law 2016). Prior to the earthquakes, in 2009, the Mayor at the time, Bob Parker, had also signed a community petition that 'called on the Council to build an aquatic recreation facility in the east' (Kenny 2021b). In addition, the Linwood ward councillor had campaigned strongly for an aquatic centre in the eastern suburbs for over a decade.

In the coastal suburb of New Brighton, He Puna Taimoana, an $11.2 million hot saltwater pool facility was opened in May 2020. Retail and hospitality businesses in New Brighton had struggled for several decades to attract visitors, despite a range of initiatives such as a weekend market. The community's somewhat precarious economic situation was exacerbated by the earthquakes, with damage to buildings and roads discouraging people from other parts of the city from visiting the area. Regenerate Christchurch, the rebuild planning agency set up in 2016, had a legal brief to work to attract further external investment to New Brighton. Given the economic challenges faced by the suburb, the decision of Development Christchurch – a city council-owned economic development agency also formed in 2016 – to invest in the new saltwater pools was significant. Prior to their opening, the manager had expressed her hopes for the wider impact of the facility, explaining that they were 'looking to get a host of visitors to come back to New Brighton and get a fresh taste of an old favourite' (Kenny 2020). Like the new Taiora complex, this facility has proved popular since its opening, and is often fully booked.

The wider economic and environmental situation for New Brighton remains challenging, however, as its low-lying coastal location leaves it vulnerable to sea-level rise. The earthquake sequence considerably exacerbated this issue. Due to liquefaction, nearly 90 per cent of central and eastern Christchurch subsided, often by half a metre, with parts of the red zone sinking by a metre or more (McDowall and Denee 2019). This renders the areas along and behind the coastline even more vulnerable to sea level rise and storm effects (Hughes et al. 2015; Monk et al. 2016) and will decisively affect the residential suitability of New Brighton in future decades. The city council will have to grapple with extended areas of managed retreat. In effect, this is what has already happened in the residential red zone, and the situation in New Brighton will likely also involve having to incentivise or compensate residents to relocate. In any case, anticipated sea-level rise will surely alter the economic and political calculus around future public investments in New Brighton, including whether they can be justified and on what basis.

Viewed more broadly, however, three of these developments – Taiora QEII, Te Pou Toetoe and He Puna Taimoana – demonstrate the responsiveness of the city council and its partner agencies to the needs of residents in the eastern suburbs. Although some of the facilities took time to build and considerable lobbying, they nevertheless reflect a substantive investment in local sporting facilities, which will contribute to the health and well-being of children, youth and adults in the east. This sensitivity to local needs contrasts with the city centre focus of central government investment in the blueprint, particularly in terms of the Crown anchor projects (Chapters 4 and 9).

Educational upheaval

Alongside sport and recreational facilities, public educational provision was significantly affected by the earthquakes, with extensive damage to many schools in the east. For example, the facilities of the main high schools serving the area – Aranui High, Shirley Boys' and Avonside Girls' – were rendered unsuitable for use by the February 2011 quake. Along with several other schools in eastern Christchurch, Shirley Boys and Avonside Girls entered into site-sharing arrangements with less damaged and still operational schools in the west of the city. The teaching day was rescheduled into two parts, with students from the host school attending from early morning to early afternoon, and students from the eastern schools then attending from mid-afternoon to early evening. This site-sharing lasted between 6 and 12 months for the schools involved.

In September 2012, the Ministry of Education announced a comprehensive set of proposed changes in the form of the Christchurch Schools Rebuild programme (Ministry of Education 2021). Scheduled to begin in 2013, the plan was to rebuild or repair no less than 115 earthquake-damaged schools in the greater Christchurch area. In addition, 13 schools were to be closed and a further 18 merged. Aranui Primary and Secondary Schools were slated for closure. Avonside Girls' was to be closed or merged with Christchurch Girls' High and Shirley Boys' was to be closed or merged with Christchurch Boys' High School (Law 2012). The closures and mergers, which disproportionately involved eastern schools, were justified with reference to the expense of earthquake repairs – the anticipated $750 million cost was neither fiscally nor politically attractive – and the westward shift in the city's population. The latter factor was particularly salient for eastern suburbs, as outmigration from red zoning and housing damage had seen falling school rolls. These falls were often used as a justification for the proposed closure or merger of particular schools.

Unsurprisingly, these proposed school closures generated strong responses from residents, children, teachers and principals in the eastern suburbs. Many communities had deep attachments to their local schools, reflecting not only their educational purpose but also their function as places for sheltering and gathering during the earthquakes. The proposals seemed to demonstrate a lack of awareness, even indifference, to the significant roles of schools within their communities (Mutch 2018). The President of the New Zealand Parent Teacher Association asked 'Hasn't Canterbury and Christchurch suffered enough? It is time to move forward and listen a little more to what parents and parent groups are saying before making the decisions that will affect our communities in such a dramatic way' (Bayer 2012). The Headmaster of Shirley Boys High School, which had a tradition of welcoming pupils from a range of social and economic backgrounds, was also fiercely opposed: 'I state here and now … Shirley Boys' High School as a school exists and will continue to exist – mark it. There is no way in God's creation that we cease to exist' (Radio New Zealand 2012).

Picking up on these community concerns, an editorial in *The Dominion Post*, the Wellington daily newspaper, condemned both the substance of the proposals and their insensitivity to local context. 'The Ministry's first cut at restructuring Christchurch schools looked like it was devised on a Wellington whiteboard without reference to the situation on the ground' (Stuff 2013).

In a similar vein, Mutch (2017, 74) argued that the Christchurch Schools Rebuild programme demonstrated 'a lack of understanding of, or even blatant disregard for, the place of schools in communities and the importance of communities in sustaining a stable and cohesive society'.

Following strong community feedback, there was a number of adjustments to the original plans. The Minister of Education announced in February 2015 that both Avonside Girls' and Shirley Boys' schools would be rebuilt as a combined campus adjacent to the new Taiora QEII sport and recreation facility. This opened in 2019. In addition, Linwood College was entirely rebuilt on a new site, opening as Te Aratai College in May 2022. These examples indicate that the voice of eastern communities could influence central government plans to some extent.

Overall, however, educational provision in the eastern suburbs was heavily impacted by the rebuild programme, with 13 schools closed. Aranui, Avondale and Wainoni primary schools and Aranui High School were shut in January 2017, for example, and a new year 1–13 integrated school, the Haeata Community Campus, opened the following month. The Minister of Education explained that Haeata would bring 'a fresh and exciting approach to education in Aranui, with huge potential benefits for students and the wider community', and that the school would be 'the first campus of its kind in Greater Christchurch, with successful examples operating in other parts of New Zealand' (*New Zealand Herald* 2013). When Haeata opened, it had a buoyant roll of 955. As an 'unzoned' school, it had been able to attract students from across the city. Just four years later, in March 2021, the roll had fallen by 40 per cent to 595 students. Students were reported to be unhappy with the open plan, multi-age classrooms and the self-directed model of learning. National examination (NCEA) attainment rates in 2018 were low, with only 27 per cent of year 12 students gaining NCEA level 2 (compared to 77 per cent nationally) and just 14 per cent of year 13 students gaining NCEA level 3 (compared to 66 per cent nationally) (O'Callaghan 2019). As NCEA level 2 is generally considered the minimum qualification in New Zealand for a young person to move on to further education, training and employment, these were problematic outcomes in an already disadvantaged area (Table 10.1). Although attainment levels may improve, the merger of four smaller schools into a single campus and the adoption of an experimental educational model is not yet working well for young people in this part of eastern Christchurch.

So, while the rebuild programme saw substantial investment in Christchurch schools as a whole, for children and families in the eastern suburbs the outcomes were clearly less favourable. An Ombudsman's investigation into the Christchurch Schools Rebuild programme identified serious shortcomings in the Ministry of Education's initial consultation process (Boshier 2017; Radio New Zealand 2013), and there was media criticism of 'the high-handed manner of the minister and her ministry and the subsequent scramble of both to make amends' (Stuff 2013). After the Ombudsman's findings, the Ministry of Education apologised for its lack of transparency and failure to engage meaningfully with the affected communities. By then, however, several schools in the eastern suburbs had already been closed, despite the protest of local residents. Coming just one year after the announcement of red zoning, the manner in which the schools were closed did little to alleviate the mistrust that some eastern residents had developed towards the government.

The forgotten east?

If the behaviour of government, business and community actors is considered across the four dimensions of recovery discussed thus far it is possible to assess the degree to which the eastern suburbs were forgotten or neglected. A number of points can be made. First, in terms of investment and rebuilding activity, the post-earthquake focus of central government was primarily on the central city, as Chapter 4 has shown. This is evident in the emphasis given to the large anchor projects such as Te Pae and the Parakiore Recreation and Sport Centre, described in Chapter 9. Such a focus is arguably characteristic of right-of-centre governments engaged in post-disaster recovery, in that a command-and-control approach is typically employed in an effort to repair buildings, replace damaged infrastructure and foster business renewal (Cretney 2017). While such a neoliberal approach can be effective for dealing with the economic and built environment aspects of recovery, if attention is not also paid to its psychosocial and community dimensions then additional disturbance and harm to communities may follow, beyond that associated with the initial disaster (Quarantelli 1999).

In Christchurch, the National-led government's focus on the central city appears to have been coupled with an unwillingness to recognise or take account of the differentiated experience of the disaster across different suburbs. This spatial insensitivity was evident in the blanket nature of the red zone, and the unyielding refusal to accommodate substantive variations in the circumstances of those subject to red zoning (eg. the uninsured, or those whose property values significantly exceeded the Crown's financial offer). A similar inflexibility was evident in the Christchurch Schools Rebuild programme, with the closures of several eastern schools for reasons that some consider to be as much ideological as financial or logical (Mutch 2017). The rebuild programme also failed to take into account the significant intertwining of schools and their communities. Such an oversight could be characterised as forgetting, although it was not clear whether the Ministry of Education staff in Wellington were ever adequately attuned to these place-based dynamics, or whether their actions were in some way particular to eastern Christchurch.

Second, in comparison to central government, the city council was more responsive to the recovery needs of east. This reflects the council's mandated responsibility to city residents, and its system of representation through ward-based councillors. But the new Taiora/QEII, Linwood and New Brighton pools were only built after thorough consideration of relevant factors, including the westward shift in the city's population, plans for the jointly funded Parakiore facility in the central city and a council already under financial strain by virtue of its contributions to the repair of hundreds of kilometres of damaged subsurface infrastructure. None of these factors was favourable to building new facilities in the eastern suburbs, and the council's decision to do so arguably reflects its recognition of, and responsiveness to, the needs of eastern residents. At the same time, the relocation of athletics facilities that had once been at QEII to the new Ngā Puna Wai facility might be considered an appropriate response to the substantial population growth in the southwest of the city.

Third, there were many instances when private insurance companies were slow to respond to the repair or rebuild claims of homeowners in the eastern suburbs. This does not appear to reflect any specific or systemic discrimination against eastern residents, but was rather a more general problem with the capacity and inclination of insurers to settle claims in a timely fashion, which affected homeowners across the entire city (Miles 2016). The problems eastern residents experienced with insurers seem to have derived primarily from the concentration of damage in their neighbourhoods (and the time required for additional geotechnical investigations to be undertaken in some cases). Some insurance delays also reflected the intertwined nature of claims, such as in the case of multiple-unit housing complexes, where the satisfactory settlement of one claim was often linked to the progress of another. But this appears to have been a city-wide issue, as Chapter 5 suggests, rather than one which reflected discrimination against eastern residents.

Fourth, with respect to private sector developments in the aftermath of the earthquakes, the great majority has occurred either in the central city, in the form of new retail and entertainment complexes, cafes, hotels and apartment buildings, or in its southwest, west and northern areas. If the affluent seaside suburbs of Redcliffs and Sumner are discounted, then the Tannery retail complex in Woolston is the only substantial private sector development in an eastern suburb since the earthquakes. Although this encompasses a significant number of independent retailers, it is undoubtedly something of an exception. So, the central city and western orientation of new commercial activity is arguably less a matter of 'forgetting the east' than of investors choosing locations they believe will generate profits and reduce risk (eg. liquefaction from possible future seismicity). In this sense, these businesses were operating as they might normally be expected to. Central government, with its responsibilities to the broader population, can, however, be judged by different standards.

Conclusion

Were the eastern suburbs of Christchurch in some way neglected or overlooked in the post-quake recovery process? In terms of the attention paid to the concerns of eastern residents, a distinction can be drawn between recovery initiatives led by central government agencies – the residential red zone and the reorganisation of schools – and the activities of local government and community groups such as WeCan. Under the leadership of a

National-led government in office from 2008 to 2017, recovery initiatives were often characterised by top-down decision making and, in several cases, an apparent indifference to the needs of particular communities and residents, despite some community consultations undertaken by CERA. The notable exception is in the area of social housing, where Housing New Zealand Corporation and then Kāinga Ora have overseen significant new developments in places such as Richmond and Aranui.

The city council was consistently more attentive to residents in the eastern suburbs. In response to local advocacy and submissions over several years, a number of new sports and aquatic facilities were built. The actions of various community groups (eg. WeCAN) ameliorated aspects of the central government recovery initiatives, such as the proposed closure of Avonside Girls' and Shirley Boys' schools and the initial lack of financial compensation for uninsured red-zoned households. In a hypothetical situation where a centre right government had been in sole charge of post-earthquake recovery in Christchurch, without the compensating influence of both the city council and community groups, the eastern suburbs would likely have fared rather worse in terms of political attention, reinvestment and rebuilding. Instead, the actions of council and community helped to ensure that the concerns of eastern residents were not only heard but to some degree also addressed. This underscores the importance of ensuring local and democratic input within the disaster recovery process, so as to achieve more socially and spatially equitable outcomes (Quarantelli 1999; Cretney 2018).

Following the election of a Labour-led government in October 2017, there was a discernible shift in the approach taken by central government towards post-earthquake recovery in Christchurch. The new Minister for Greater Christchurch Regeneration took a more proactive stance when addressing unresolved Earthquake Commission claims, as well recognising the high number of cases where the initial work undertaken was in need of subsequent repair. This was the reason for the establishment of the Canterbury Earthquakes Insurance Tribunal in 2019. And with the acceptance by the Minister of the Ōtākaro Avon Regeneration Plan in 2019, and the Global Settlement in 2020 that promised the return of red-zoned land to the city, the stage was set for the creation of more positive stories and experiences in eastern Christchurch. These are explored in the next two chapters.

11 The more-than-human city

Introduction

The previous chapter has presented a clear picture of differentiated social experience of post-earthquake recovery in the eastern suburbs of Christchurch. However, a post-disaster city is a theatre involving both a cast of multiple kinds of actors and a plethora of diverse but inter-related kinds of performances. If the Canterbury earthquakes achieved one thing at the time, it was to disabuse its inhabitants and those of surrounding areas of a foundational belief of western urban culture: that all of nature is a fixed stage upon which to enact the human drama. These profoundly unsettling times disturbed the anthropocentric imaginaries of city spaces and politics built upon the presumption of solid ground. Such imaginaries typified the nineteenth-century colonial project of which the city was a product, the goal of which was an ongoing process of terraforming, or 'improvement' without limits (Ghosh 2021; Latour 2018). Improvement depended on the conversion of earthly resources to create both value in urban form and the sale of commodities to support that form. Yet the shaking of the ground and its widespread liquefaction revealed the extent to which the chimera of improvement had always been conditional. The rhythms of the deep times of nature have no regard for the social (Irvine 2017).

As a self-proclaimed 'garden city', however, Christchurch has long considered itself as owning natural attributes defined – as befits the garden metaphor – in human rather than more-than-human terms (Pawson 1999, 2000). It has also been a place long seeking to capitalise on the 'stunning natural landscapes' of rivers, hills and coastlines, all of which are performatively enrolled in a much-vaunted 'active lifestyle' (Christchurch NZ 2021). But the earthquakes and their aftermath have brought home the extent to which the city is a hybrid assemblage of unstable ground, surface and subsurface waters, unruly flora and fauna and much else, in addition to its fixed attributes of roads, pipes, commercial buildings and parks, and the affective qualities of homes, neighbourhoods and landmarks.

This chapter uses the concept of the 'more-than-human' to explore the importance of interconnections between human and non-human factors in the the city in transition. As such it focuses on how actual worlds become assembled

DOI: 10.4324/9780429275562-14

when all manner of things (or 'actants') make combined contributions to the making of particular geographies. In so doing, it acknowledges how a variety of non-human actants – species, forces, materials, contexts and discourses – all interact with human behaviour and decision-making to produce worlds that are more-than-human. And it assesses the extent to which opportunities for more lively spaces, hybrid ecosystems and species interactions are being recognised.

The 'more-than-human' city

In recent decades, social scientists have become discontented with the tendency to understand nature spaces in binary terms (Cloke and Johnston 2005). Rather than using mutually exclusive categories such as 'human' and 'non-human' or 'nature' and 'culture', researchers have focussed on ideas about relational or hybrid agency (Whatmore 2002) that demonstrate how interactions and inter-reliance between these categories reveal a surplus of meaning and understanding – a sum total that is 'more than' its parts. In this way, previous attention to 'human' agency, and (to a lesser extent) 'non-human' agency, is being superseded by a concern for the excess that spills out when these categories are understood relationally – that is, according to their inter-relations rather than any particular innate qualities. The early fruits of this approach are visible in the growth of animal geographies (Buller 2014) that trace the intersection of animals and human society, demonstrating how animals can transgress and help to reshape the imagined spatial and social ordering of that society. More-than-human perspectives have also been used to understand the relational contribution of trees to places (Jones and Cloke 2002), for example, in providing capacity for culturally significant memorialisation of key people and events (Cloke and Pawson 2008).

Adopting a more-than-human approach has enabled new lines of enquiry and action to be developed which have challenged the anthropocentric status quo in at least three ways. First, in an emphasis on entanglement, there is the realisation that humans and non-humans co-constitute each other; that is human existence and social life depend on often asymmetrical relations with non-human bodies and material things (Hodder 2014). It follows that any presumed dominance of human issues needs to be de-centred in recognition of the myriad ways in which non-human actants are entrenched in and intertwined with social, cultural, political and infrastructural dimensions of life (Bennett 2010). Second, such entanglement involves a liveliness that cannot easily be shrugged off by assuming that humans are active and non-humans are passive. Indeed, non-human contributions to relational agency have been recognised as unpredictable, energetic, transformational and full of unseen capacity and potential (Clark 2011). Such contributions have been labelled as 'vitalist', emphasising the liveliness of the non-human – such as seismic activity – in helping to animate and agitate human relations (Greenhough 2010).

Third, more-than-human approaches underscore the need for entangled empathy (Gruen 2015), in which the de-centring of human agency is accompanied by an equivalent displacement of humans as the defining reference point for ethical relations. As Cloke and Jones (2003) argue, the tightly drawn boundaries which codify moral communities as human need to be reconceived to foster new and more care-full relationships which are attentive to more-than-human nature. Mere recognition of a more-than-human world is insufficient – empathetic care for that more-than-human world needs to be practised in a polity with more channels of communication between its constituent parts (Puig de la Bellacasa 2017). In short, a more-than-human understanding of the world requires the dragging of everyday political practices of society into new ethical formations (Stengers 2005).

These ideas have been translated into practical urban contexts in accounts of the more-than-human city. Far from a blank canvas on which anthropocentric designs are inscribed, cities are lively places characterised by hybrid ecosystems where human lives are entangled with those of myriad non-humans. Cities are 'overflowing, exuberant, excessive and phantasmogoric' (Franklin 2017, 203), teeming with heterogeneous forms of life capable of disrupting and reconstituting seemingly stable histories, narratives and materialities. More-than-human liveliness occurs at varying scales. Clark (2011) emphasises how cities are vulnerable to planetary forces such as those relating to climate change and seismicity that resist anthropogenic control and planning, and replace human-centred polity with turbulent unpredictability. Even at a localised scale, urban order can be disorganised and disarranged by the power of more-than-human life, like tremors, fire and flood, thus reconstituting assumed or desired norms.

This picture of the more-than-human city is particularly relevant to Christchurch, where planetary forces have been evident in the event of the earthquakes, and where subsequent planning and place-making have been deployed to re-create life in the city. As Chugh (2020) explains, urban design often follows a logic dictated by human-centred imagination and spatial ordering, in which non-human nature tends to be controlled using tactics of pre-assembly and segregation. The capacity of the non-human to repurpose lively infrastructure and to furnish novel ecologies can be forgotten in amongst the blueprinting of zones and precincts. Such modernist visions (discussed in Chapter 4) are often assumed to be enacted on a neutral substrate, but in terms of the more-than-human city, that substrate needs to be seen as active, lively, fragile and relational in ways that matter.

The rest of this chapter uses these ideas to interrogate aspects of the emerging post-disaster city of Christchurch. Following Franklin (2017), it addresses the ways environmental and technical forces have become incorporated into the life, subjectivities and structures of Christchurch after the earthquakes. It illustrates how particular relational actants have become significant companions to city dwellers whose lives have been ruptured by geophysical forces, and the extent to which they have become key signifiers of contested spaces in the city. It then interprets the degree to which relational actants shape and embody the values and concerns of particular groups of citizens and the ways they personify and provide focus for otherwise tricky political and ethical issues in the emerging city.

Disturbing the entrenchment

The pre-earthquake city was an intricate web of relations between the human and non-human. One of the key processes of improvement was the draining of the wetlands underlying much of the city centre and inner suburbs (Schrader 1997; Wilson 1989). Few people would know that the urban area sits on the active margins of the Waimakariri River delta, on land only a few thousand years old, or that its networks of streams are sourced in groundwater bubbling to the surface across that delta. Stream courses lie interred beneath the city centre (Lucas 2014), and have often been constrained in the suburbs by timber boxing. Only since the 1990s have many been 'liberated' through more naturalistic design, their watercourses replanted in native wetland species. Such moves were initially controversial with a citizenry familiar with 'Eurocentric regiments of tall, erect, fastigiate tree forms' and 'the received fashion for manicured, sprayed and sanitised gardens and lawns' (Meurk 2021, 159). In fact, the garden city's self-image was always partial: the extent of canopy tree cover is only 11.6 per cent if large-scale plantation forests are excluded (Morgenroth 2017). Trees in parts of the city were directly affected by soil movement, liquefaction and rockfall generated by the earthquakes, but immediate tree losses were relatively low, barely 50 per cent more than the number removed on public lands in any one year (Morgenroth and Armstrong 2012). Longer-term losses have only become apparent in the decade since, and the most identifiable cause of these is housing intensification (discussed in Chapter 5).

Non-human actants were therefore intricately and long-entrenched in the social, political, historical and infrastructural life of the city. These entanglements were disturbed in a variety of ways during and after the event of the earthquakes. Many of the immediate disturbances have been well documented (Environment Canterbury 2013; Potter et al. 2015). For example, the incidence of liquefaction discussed in Chapter 2 resulted in large volumes of silt being deposited into the city's stream courses, resulting in significant increases in suspended sediment and detrimental changes to water quality. In turn, there was a severe impact on aquatic life, notably on sensitive fish species and shellfish, but also due to disruption of marine mammal habitats. Moreover, the first years after the earthquakes saw a sharp rise in bird mortality, often connected to changes in the chemistry of water treatment ponds but also due to spoilation of nesting and roosting sites. The inaccessibility of wider landscape habitats heralded a reduced capacity to control what were previously perceived as weeds and pest species. In such ways, the earlier entrenched relations between humans and non-humans were (literally) shaken by the earthquakes.

The impact of these kinds of disturbances on the more-than-human city can be understood in three main ways. First, many of the relations that characterised the city prior to the earthquakes endured. Companion animals, for example, continued to be important to human health and well-being in the city. Animal welfare staff and other carers looked after thousands of animals who were injured or rendered homeless during the quakes (Potts and Gadenne 2014). Several organisations – for example, redzonecats – sprang up to look after the interests of cats that were abandoned to fend for themselves during the forced migration of red zone households (redzonecats 2021). While for some, 'feral' cat activity represented a threat to indigenous wildlife, other accounts emphasise not only the human-centric values of care for animals, but also the reciprocal relations between animals and humans. Thus Coombs et al. (2015) analyse the mental, physical and social support received by dog owners through the companionship of their pets. These relations were re-prioritised with greater time and attention being given to such companionship during the earthquakes, but with any loss of togetherness exacerbating the negative impacts of the event. Heroic stories of individual quake dogs (Sessions and Bullock 2013) and quake cats (Bullock 2014), describing their exploits in situations of rescue, reunion, affection and caring support of humans, became a symbol of the resilience of the city in responding to seismic disaster.

The enduring relations of the more-than-human city were also in evidence in the reliance on green and blue landscapes for recreational well-being. One of the initial impacts of the earthquakes was to restrict opportunities for those outdoor activities such as tramping (hiking), running and cycling that are a fundamental part of New Zealand culture. Access to the Port Hills adjacent to the city was initially severely curtailed, with more than 50 kilometres of paths and tracks closed due to rock fall hazards (Thompson et al. 2015). Water-related activities like rowing, kayaking and sailing were also significantly hampered (Xayasenh 2015). However, residents continued to seek outdoor recreation where they could, largely in the city's undamaged parks and green spaces, in a lasting collaboration between green landscapes and well-being. The Botanic Gardens, long a site of city recreation and memories, became a popular haunt as one of the few undamaged sites in the city centre. As in New Orleans after Hurricane Katrina (Tidball 2014), trees acted as beacons of hope, survival, stability and place in the process of recovery from disaster, just as thousands of refugees from city hotels and apartments had sheltered beneath the trees of Hagley Park on the night of the major quake of 22 February 2011.

A second way of understanding the impact of the earthquakes is to identify the ways in which the cultures and practicalities afforded by the quakes became incorporated into everyday life in the more-than-human city. A good example are the changes wrought on the regulatory and architectural environments of the rebuilt city, prompting radical changes to the materiality of the cityscape, and to the lived relations of co-existence within it. Here, the technological adjustment required to proof the city against future seismic risk has brought its own liveliness – functional and aesthetic – to its look and feel. Initially, such adjustments led to the very obvious presence of shipping containers in the urban environment, both as functional props for vulnerable sagging buildings, and as defensive walls against rockfall around the coastal suburb of Sumner. They were also influential elements in emergent transitional architecture, whether as building blocks for the temporary container mall, or as foundational anchors for Shigaru Ban's acclaimed 'cardboard cathedral', a larger version of one designed by the architect after the Kobe earthquake in Japan in 1995 (Walsh 2020). The strangely out-of-place materiality of the containers became incorporated into the fabric of the city (Figure 11.1).

Figure 11.1 Post-earthquake uses of shipping containers in the urban environment: (top) the 'wall of Sumner', January 2012, since removed; and (bottom) Tuam Street, central city, March 2022.

As rebuilding took on a longer-term perspective, the technological regulation of earthquake-proofing took on different forms. The city centre's architecture – dominated by unreinforced masonry and concrete buildings – became obsolete in the post-quake setting where the inelastic failure of these materials to resist tensile forces was recognised in new seismic building codes and health and safety regulations (Gjerde 2017a). These codes and regulations created buildings exhibiting 'structural stoutness', visible structural cross-bracing and a steel and glass aesthetic (Gjerde 2017b). Simultaneously, damping devices, energy dissipators and base isolators provided unseen but vital insurance against the repeat of disaster, replicating those in the city's casino built in the 1990s, and its Women's Hospital (opened in 2005), the only pre-earthquake buildings in Christchurch so protected. In these ways, transformational technologies, materialities and narratives born of the earthquakes became incorporated into entanglements of the more-than-human city.

A third aspect of the entanglements of the more-than-human city after the earthquakes relates to the way in which relational actants come to signify contested spaces. Franklin (2017) notes how such actants form and embody the principles, values and interests of different groups of citizens and how they become a focus and 'voice' of otherwise difficult political and ethical values in urban design, planning and decision-making. For example, the earthquakes provided ample scope for adventitious colonisation of urban spaces such as demolition sites and empty house sections, but not everyone welcomed the appearance of rabbits or weedy species. Others however felt encouraged to subvert conventional ways of being in the city. A specific example from the transitional movement (which was discussed in Chapter 8) is Liv Worsnop's Plant Gang, an environmental art practice of planting succulents and other rubble-ready species in empty city centre sites (Boswell 2021). It is echoed in her etchings of weeds of the rebuild on the large glass windows of the re-located C1 Espresso café, one of the first social institutions to re-open as the city centre cordon shrank. Its new home in a repurposed post office had been erected to high seismic standards after the 1931 Hawkes Bay earthquake but remained surrounded by ruins and abandoned sites for several years after re-opening in 2012.

Two narratives of contest

Two sites in the city that became notable, in different ways, for contests between human and more-than-human vitalities are the residential red zone and the abandoned, flooded basement of a demolished city centre tower block. In both cases, the disturbance was the result of the manner in which bird species have responded to the availability of new post-earthquake habitats. In the residential red zone, whose creation and future are discussed in Chapters 10 and 12, Crown agencies sought to ensure that interventions like those of the Plant Gang would be discouraged. As the Canterbury Earthquake Recovery Authority (CERA) cleared house sites, it left in place 'whole suburbs constructed only with vegetation' (Bowring 2021, 111). This is a reference to those garden trees and shrubs that after initial public outcry it was persuaded to leave behind. There were a lot: CERA compiled a database of nearly 30,000, about 1800 of which were fruit and nut trees (Gates 2015c). But around these trees, the land was levelled and sown in grass, which was then regularly mown. After the dissolution of CERA in 2016, Land Information New Zealand, the manager of Crown lands nationally, continued this policy of close mowing, in effect creating a parkland landscape entirely consistent with the city's 'civilising' ethos of improvement.

Although community groups such as Greening the Red Zone recorded the very limited patches of native vegetation in the Ōtākaro Avon river corridor and were permitted to extend some of these with public plantings of indigenous flora, the Crown land management policy produced a double irony. First, it set back the rewilding that is the centrepiece of the widely supported regeneration plan (discussed in Chapter 12) by nearly a decade. Secondly using mowers to minimise the risk of encouraging populations of rats and mice created ideal habitats for other species. One such is Canada geese (*Branta canadensis*), which are attracted to the short pasture habitats and wetlands that abound in the red zone. They have also been encouraged by the provision of managed wetlands designed as flood mitigation devices in many post-earthquake greenfield suburbs (Chapter 5) as well as to the irrigated pastures of dairy farms, rural Canterbury's twenty-first-century boom industry (Le Heron 2018). The birds lost their game status a decade ago, so are no longer managed by Fish and Game North Canterbury, the regional agency responsible for sports fish and game bird populations. They have subsequently spread into these new urban and rural niches.

The city population of Canada geese reached about 5000 individuals, bringing attendant problems of water contamination, intra-species competition with other water birds and fouling of public places and pathways (McDonald 2020e). Nonetheless, Land Information New Zealand (2021) claimed that its 'vigilant and pre-emptive management approach' has kept 'pest species' in the red zones at acceptable levels, although recent survey data has focused only on small mammals, namely mice, rodents and mustelids. Rats and mustelids have been controlled by trapping networks monitored by community volunteers. Rabbits, which also thrive in short grass environments, have long been regarded as a threat to New Zealand's pastoral industries, and as such are subject to the Canterbury Regional Pest Management Strategy. When red zone rabbit populations have reached 'non-compliant' levels, they have been controlled by Rabbit Haemorrhagic Virus Disease administered as a biocide (Boffa Miskell 2020). In 1997, when this disease was surreptitiously first introduced into New Zealand, there was some concern about its cruel effects, but its use in the red zone has not been sufficiently publicised to arouse public emotion.

A second narrative of contestation in the more-than-human city relates to the open basement of a demolished office tower on Armagh Street in the central city, then owned by the Carter Group. Hinchliffe and Whatmore (2006) have demonstrated that the arrival of a new species into urban habitats results in dances of agency which invoke new assemblages of constraint and opportunity. These can provoke contested politics involving both denaturalisation and attempts to provide empathetic care. In 2018, a flock of some 300 black-billed gulls (tarāpuka in te reo Māori) arrived at the site, nesting in the concrete beams and flooded foundation stumps of the former building. The tarāpuka is the most threatened of New Zealand's three indigenous gull species, and is regarded by nature-conscious people as both precious and rare. The species breeds mostly in the braided river beds of Canterbury and Southland, and is threatened by predators, habitat loss and human disturbance. The cordoned-off site in central Christchurch offered both safety from predation and disturbance, and a physical environment suited to nest-building (Figure 11.2).

The arrival of the tarāpuka in the post-earthquake city provoked a series of contested discourses and political responses. The Department of Conservation took an immediate interest because the Wildlife Act 1953 dictates that once the birds start nesting, it is illegal to disturb them until the end of the nesting season. The specification of the tarāpuka as 'near threatened' ensured that conservation wardens monitored the site, even designing and installing purpose-built floating platforms as life rafts for chicks that were falling from their nests into the flooded basement of the building (Hunt 2021b). However, departmental representatives made it clear that the city centre was not an ideal place for the tarāpuka and that the birds should be encouraged to move back to their more 'natural' breeding sites (Gates 2021). Meanwhile, the charismatic nature of the tarāpuka, along with the publicity given to how their colony was flourishing in their new urban environment, ensured that they became something of a zoological attraction for residents and tourists grasping an opportunity to learn about these rare native birds at close quarters. Some even suggested that the tarāpuka represented a positive adornment of a site that was otherwise one of the city's ugliest post-quake eyesores (Bowron 2019).

To the site's owners, however, the tarāpuka represented a significant hindrance to their capacity to redevelop. This frustration heightened with an agreement in 2019 to sell part of the land for the construction of a new Catholic cathedral and to use the remainder in a joint venture to build an associated pastoral hub, primary school, offices, residential accommodation for clergy and multistorey car parking. The development was hailed as an important piece of the central city chessboard of redevelopment (Blundell 2021), linking Te Papa Ōtākaro (the river precinct), the performing arts precinct and Te Pae, the new convention centre (McDonald 2019d) (Chapter 9). A new alignment of interested parties thus emerged, to which owners of surrounding hospitality businesses were soon added due to the nuisance caused by the tarāpuka in harassing diners, scavenging food and fouling buildings. In this way, the legislative safety net of care, and a potential politics of entangled empathy, became quickly outweighed by economically

Figure 11.2 Tarāpuka colonising a central city site.

Credit: https://annettewoodford.wordpress.com/

driven positioning of the tarāpuka as 'unnatural' urban residents out of place in the post-disaster city. A plan was agreed for a gentle eviction of the birds at the end of the 2019 breeding season using nets, with spikes to deter them from returning.

This apparently uneven contest was, however, waylaid by the lively agency of the birds themselves. The installation of netting at the site was not completed before the tarāpuka came back to nest in 2020. A further year of occupation of the site ensued, during which the presence of the birds gave energy to longer-term arguments about rewilding the city and the capacity of indigenous species to provide local flavour and natural character (cf. Meurk and Swaffield 2000). Having failed with the eviction the year before, developers transformed the site in 2021 with the recycling of the ruined foundations to make way for the parking building. The beams and pillars previously used as nesting sites were torn down prior to the breeding season, and liaison with local property owners resulted in plans to prevent birds from congregating and nesting on nearby rooflines. The tarāpuka scattered along the Ōtākaro Avon River and into the residential red zone, only to return to the site when demolition was suspended in August 2021 due to the imposition of a Covid-19 level 4 lockdown in Christchurch. Meanwhile, an entire breeding colony of tarāpuka abandoned its nests in the Upper Waimakariri riverbed due to a suspected dog attack (Allot 2021). Ironically, breeding conditions at this 'natural' site were less safe than in the supposedly 'unnatural' Armagh Street venue. Subsequently, some tarāpuka found a new home in a partially constructed stormwater basin in the suburb of Belfast. Here, according to Law (2022), as workers downed tools during the Covid-19 lockdown, the birds appeared to mistake the basin site for a braided river and began nesting again in an 'unnatural' urban location.

The new dances of agency performed by actants such as tarāpuka and Canada geese certainly brought about material responses and a questioning of the place of non-human actants within and beyond the redevelopment of the central city. In Chapter 8, it was argued that the importance of transitional gap-filling and greening initiatives after the earthquakes became acknowledged in longer-term re-evaluations of the politics of placemaking so as to include community orientation and playful theatricality amongst commercially-driven development. In a similarly transitional manner, it can be suggested that the borrowing of urban sites by tarāpuka and Canada geese may find longer-term significance in the recognition of the more-than-human city as a complex configuration of landscapes and a mosaic of ecosystems. Inevitably such a vision needs both to disturb any rigid politics of how non-human actants are 'in' or 'out of place', and to recognise that the relational agencies concerned are unpredictable and subject to continual change. Such characteristics of the more-than-human city also involve the excitement of new experimental opportunities for entanglement, and it is to this possibility that the chapter now turns.

Emerging entanglements

The rebuilding of Christchurch after the earthquakes has offered up a series of opportunities to develop new experimental configurations of more-than-human relations in the city. In general, such reconfigurations require a recognition of the need for political change that in some way enables, or at least makes room for, a disturbance of conventional political wisdom and assumptions about nature-society relations. In this case, it was the rupture of the earthquakes that created different kinds of potential spaces in which alternative ideas, narratives and discourses could find expression. At various scales, material, cultural and performative, opportunities arose to explore in particular ways new forms of more-than-human entanglement and thereby to try out alternative ideascapes, practices and even moralities capable of co-constituting city worlds differently.

One of the most important opportunities offered by the rebuild has been a more overt recognition of indigenous entanglements with land and water, a theme introduced in a number of preceding chapters. The Canterbury Earthquake Recovery Act 2011 required that Te Rūnanga o Ngāi Tahu be consulted in the development of a recovery strategy and 'have the opportunity to provide input' into the Central City Recovery Plan (the blueprint). The rūnanga (or tribal council) was also recognised under the Greater Christchurch Regeneration Act 2016 as a statutory partner to the Crown, regional and local government agencies in post-earthquake redevelopment. One response to this was the exposition of Ngāi Tahu's 'grand narratives' of living in the landscape. The aim of the document is for 'the two Treaty partners to be woven into the national narrative of Christchurch' (Tau 2016, 31). In a critique of the modernist assumptions of the blueprint, it asserts that

> it is apparent that the design teams have very little understanding of things Māori. It also appears that the designers have little knowledge of Christchurch's European heritage or culture and this does concern us – as we wonder how the design teams will incorporate and interpret European history, let alone Ngāi Tahu history.
>
> (Tau 2016, 11)

In the central city, this concern prompted the establishment of the Matapopore Trust, whose mission is to ensure 'Ngāi Tuāhuriri/Ngāi Tahu values, aspirations and narratives are realised within the recovery of Christchurch'. Earlier chapters have outlined the implications of this for the central city anchor projects. But throughout the urban area, a key concern of the 'grand narratives' is the restoration of values of mahinga kai, or 'all food producing places' (Tau 2016, 46, 48). Repeatedly it is stated that 'this does not mean that the city's 'English character' needs to be downplayed or forgotten' (Tau 2016, 60), despite that character being the product of the very processes of improvement that led to widespread destruction of mahinga kai values to begin with. For example, in the Ōtākaro Avon river corridor, this entails attention to improving water quality and ecology, potentially issues of benefit to all (Regenerate Christchurch 2019). How this will play out in practice depends on the role of the new co-governance entity that will be responsible for the future care of this area, discussed in Chapter 12. The culturally and recreationally important fishery for inanga and other whitebait species is central to this. Reversing the decline of this fishery lies in understanding the spawning habitats of the fish, and anthropogenic threats to these (Orchard et al. 2018).

Another significant area of opportunity for the development of new forms of more-than-human entanglement is evident in the discourses and proposals emanating from advocates of new landscape urbanism in the city (Meurk and Swaffield 2007). Their concern is that the urban population has become distanced from the ecological processes of its surrounding environment, being increasingly influenced by artificial simulacra of nature in for example video clips and gaming, rather than more grounded understandings of natural systems. Dealing with this perceived experiential decoupling is seen to involve two manoeuvres, both of which were conceived before the earthquakes, but became more practicable in its aftermath. First, there needs to be an appealing discursive tag with associated storytelling to draw public attention to and earn approval of renegotiated features of landscape urbanism. Second, the current patchwork of blue and green spaces and corridors in and around the city has to be configured both to promote ecological and cultural sustainability of indigenous nature and to ensure some connectivity of habitats for wildlife, thus rendering nature more visible for residents (Figure 11.3).

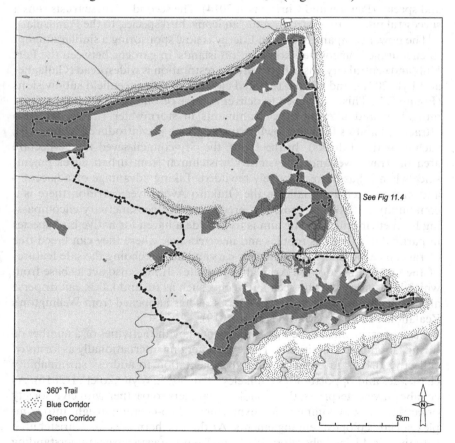

Figure 11.3 Locations of Christchurch's blue-green corridors and the 360 Trail.

Source: Trail route from https://christchurch360trail.org.nz

These manoeuvres are demonstrated in the form of an initiative to earn Christchurch the title of a 'national park city'. This was energised in 2021 in a campaign led by *The Press*, the metropolitan daily newspaper, potentially collecting under one umbrella a range of projects concerning nature-based recreation, amenity values and rewilding. It is a fresh interpretation of the rather tired 'garden city' moniker, focusing on connecting the region's blue and green corridors, so mobilising them as means of encouraging new human and more-than-human encounters. There are regional precedents for this: for example Te Ara Kakariki, the Greenway Canterbury Trust, whose object is to create a corridor of native biodiversity 'green dots' between the Waimakariri and Rakaia rivers, linking the mountains to the sea and Te Waihora/Lake Ellesmere. The Green Dots programme works with public and private landowners on the Canterbury Plains. In Christchurch City's large, rural Banks Peninsula ward, two other organisations, one national (the Queen Elizabeth II National Trust), and one local, initiated by landowners and farmers (the Banks Peninsula Conservation Trust), work to preserve and enhance indigenous landscapes and species (Pawson and Christensen 2014). The second of these trusts runs a successful project to reintroduce tūī, an iconic bird species, to the Peninsula.

The power company Meridian Energy is now sponsoring a similar attempt to encourage homeowners to plant 'tūī islands' in gardens between the Port Hills and central city. Elsewhere wetland restoration is widespread (Gallagher and Pyle 2021), and often a required element of new greenfield subdivisions (Figure 5.4). This is designed to deliver a range of outcomes, including flood mitigation and filtering of contaminants in stormwater runoff, but also attractive habitats for waterfowl (many of which are introduced game birds, such as mallard ducks). In the 1990s, the city council saved the 80-hectare area of Travis wetland in eastern Christchurch from urban development, since when it has been effectively rewilded. Taking advantage of adjacent – and higher – red zone land in the Ōtākaro Avon river corridor, there is a community initiative to provide a predator-proof ecosanctuary encompassing both environments. The aim is to provide a haven for native bird species in particular, as well as reptiles and invertebrates, where they can breed free of rats, mice, possums and domestic cats and dogs (echoing the safe features of the tarāpuka's adopted Armagh Street site). It may also act as base from which birds not presently seen in the city, such as tūī and kākā, can disperse along the city's blue and green corridors, as has happened from Wellington's Zealandia sanctuary (Marques et al. 2019).

Such experiments are being complemented by the activities of a number of 'living laboratories' in the city. These are emerging internationally as forms of collective urban governance and experimentation to address sustainability challenges and opportunities (Bulkeley et al. 2016; Voytenko et al. 2016). As collaborative enterprises, they enable researchers to partner with civil society organisations and communities, exploring and co-creating insight into complex social-ecological entanglements. At the northern edge of Christchurch City, the Styx Living Laboratory Trust aims to encourage a greater understanding of one of the city's lesser-known blue-green corridors. In the Ōtākaro Avon river corridor, one of the largest areas of managed retreat in an urban environment in the world, the Ōtākaro Living Laboratory has been established, with active support from a range of educational and governmental institutions. Its purpose is to implement one of the key objectives of the Ōtākaro Avon River Corridor Regeneration Plan, for learning, experimentation and research, 'testing and creating new ideas and ways of living' (Regenerate Christchurch 2019, 22).

This type of ambition can however disguise disparity and disagreement when projects draw on a range of discursive understandings and values about the intersection of the human with the more-than-human. A good example is the ongoing contest between the Christchurch 360 Trail and the Avon-Heathcote Estuary Ihutai Trust about access to part of the estuary edge. The 360 Trail is a 130-kilometre-long circuit of the perimeter of the city, designed to showcase its diverse contexts and ecologies (Figure 11.3). These include the exposed volcanic ridgeline of the Port Hills, the sand dunes of the New Brighton shoreline, the wetlands around Brooklands, and the braids of the Waimakariri River, through to dry grassland ecosystems beyond the airport, and residual bush remnants such as Pūtaringamotu in Riccarton. It was first proposed as the Great Perimeter Walkway in 1991, and officially opened in 2015 with support from the city council and local Rotary clubs. The ambition is in line with the Department of Conservation's Great Walks in New Zealand's national parks; the intent is to enable people to engage with the more-than-human world on their doorstep. The trail however remains only about nine-tenths complete. Some 15 kilometres on the western side of the airport cannot be officially marked, as there are no footpaths across private farmland, nor alongside roads where speed limits exceed 50 km/h.

There is also a 'missing link' around part of the Avon-Heathcote estuary in the east of the city. The estuary and surrounding wetlands are considered to be of national importance for resident and migratory birds, such as godwits, oystercatchers, cormorants, ducks, geese and swans (Marsden and Knox 2008, 759). The proximity of the city's waste treatment ponds and surrounding paddocks add to the variety of bird habitats. The part of the estuary fringe alongside the treatment ponds is a wildlife refuge, closed to the public. Companion dogs are banned from the path leading towards the refuge, the sensitivity of this issue indicated by prominent notices threatening owners with fines ranging from $300 to $100,000, or a year's imprisonment. The closed section of the estuary edge is between two to three kilometres in length and includes an access way for council staff (Figure 11.4). Alternative routes are lengthy and along major or suburban highways, taking the 360 Trail 'far from the ecologically interesting wildlife refuge' (Christchurch 360 Trail 2022). Although the city council has shown some interest in opening this link as part of its Estuary Green Edge Project, the Avon-Heathcote Estuary Ihutai Trust is resolutely opposed.

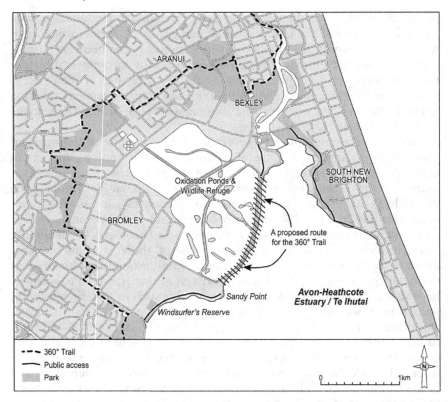

Figure 11.4 The contest between the Christchurch 360 Trail and the Avon-Heathcote
       Estuary Ihutai Trust.

The Estuary Trust was formed in 2002, with support of both the city and
regional councils. Its purpose is 'the preservation' of the estuary's 'natural
and historic resources to maintain their intrinsic values', as well as to provide
for their 'appreciation and recreational enjoyment by present and future generations' (Avon-Heathcote Estuary Ihutai Trust 2022). Over time the Trust
has supported public access to the estuary edge as well as seeking to protect
bird feeding, breeding and roosting grounds. The contentious area is the only
part of the estuary without public access, and proposals to fence the potential route along with existing dog control measures have brought no resolution. The geographical imaginaries of the two groups are very different: one
sees a short, missing link in a city-wide 'great walk'; the other considers that
as public access is available to the rest of the estuary, this area deserves protection. While the 360 Trail is 'trying to connect people to nature', the estuary
trust is committed to making 'room for nature' (Harvie 2021). At the root of
a long-running dispute are different perspectives on empathetic care, in turn
reflecting quite distinct positions on appropriate relations between the human
and more-than-human worlds.

Conclusion

In her book *Staying With The Trouble,* Donna Haraway (2016) argues for a radical configuration of relations to the earth and all its inhabitants. Recognising the Anthropocene as a self-destructive epoch of human dominance, she urges that new ways be found for a more-than-human era of living with each other. This recognises that human and non-humans are inextricably linked in practices of living and becoming together, and will require a new cultural politics of response-ability. As Latour (2018) and Dodds (2020) emphasise, the Anthropocene is not a neutral starting point for this reconfiguration. Current ecological regimes remain an essential part of late capitalism and colonialism in which ecologies are engineered and hierarchies of improvement inform disruption to human and non-human populations. To persist simply risks further degradation of relations between the human and the non-human at every imaginable scale. Haraway's plea is therefore for new expressions of staying with the earth's troubles that are conducive to the building of more liveable futures.

The realisation of Haraway's vision depends on finding a moral and political capacity to open up possibilities of becoming different. It embraces the settling of troubled waters and the rebuilding of quiet places but also necessitates the stirring up of a potent response to devastating events. It envisages the encouragement of 'collectively-producing systems that do not have self-defined or temporal boundaries ... systems [that] are evolutionary and have the potential for surprising change' (Haraway 2016, 61). As a practical starting point, she recognises the potency of developing new practices of more-than-human entanglement, thinking, storytelling and action – as well as experimental more-than-human contact zones where these practices can find expression. This chapter has suggested that a feature of the post-quake period of rebuilding in Christchurch is the nascent opening of opportunities to establish experimental sites and practices serving as pilot projects. In no small measure these opportunities have been enabled both by the material disruption of the built environment in the city centre and in the Ōtākaro Avon river corridor, and by the political rupture caused by the earthquakes within which there have been fruitful opportunities for transitional experimentation. In each case, the event of the earthquakes has opened spaces for voices and plans that were previously largely drowned out by the mainstream dialogues of the city.

The alliance of voices of activists from art, theatre and ecology with those local communities to produce experimentation in transitional place-making, leading to the formation of new discourses and partnerships in the city was illustrated in Chapter 8. This current chapter also suggests an amplification of previously marginalised voices, and emphasises, for example, the importance of Māori ideascapes and the passionate plans of landscape ecologists and other researchers in promoting an alternative polity that is better attuned to the more-than-human world. These illustrations may be understood as small steps towards Haraway's vision, but they nevertheless demonstrate a clear recognition of and empathy for the entanglements and liveliness of the more-than-human city. They may not have dragged everyday political practices into new ethical formations, but new forms of co-governance have emerged and new possibilities of more-than-human becoming have certainly been opened out.

12 The residential red zone
The city's field of dreams?

Introduction

A few weeks before the 2017 general election, *The Press* newspaper carried an editorial headed 'Unleashing excitement in the [residential] red zone will make it Christchurch's field of dreams' (*The Press* 2017). Over six years had passed since the earthquake of 22 February 2011 that destroyed the city centre and upended many suburban households, and nearly seven since the start of the Canterbury earthquake sequence. The memory of the Canterbury Earthquake Recovery Authority's (CERA's) panoptical planning, and clearance of much of the downtown to make way for its blueprint, was still fresh. There was a widespread feeling that progress in the rebuild was slow: as outlined in Chapter 4, few of the anchor projects had been completed, and some of the larger ones not even started. Many households were still embroiled in insurance disputes or awaiting repairs. The enthusiasm of Share an Idea in 2011 (Carlton 2013), when citizens swooped on the chance to express their visions for the future of the city, seemed far away. And there was an emerging view that the rebuild was producing 'the city of yesterday' (McCrone 2017). Or as the editorial put it, instead of creating 'the first new city of the 21st century ... we largely rushed to build the last version of a 20th century city ever to be constructed' (*The Press* 2017; cf. Dann 2021).

The term 'red zone' was used in a variety of ways during and after the earthquake years. The area of devastation within the central city was described as 'the red zone' whilst the cordon was in place (Chapter 6). There were also 'residential red zones', those parts of the suburbs which were so badly damaged that the Crown made offers to buy out all those homeowners who wished to move (as described in Chapter 10). And as the central city cordon shrank (Figure 6.1), within Christchurch 'the red zone' entered popular usage to describe the Ōtākaro Avon river corridor. This 11-kilometre finger of land destabilised by liquefaction and lateral spread stretches from the city centre to the sea (Figure 12.1). Setting aside other, much smaller, red zones for a while, it was this corridor that began to seem like the last opportunity to try something fresh. This chapter explores what 'something fresh' might mean, drawing on the literature about urban futures. It then describes the process of river corridor regeneration planning, before considering difficult issues of governance that are central to achieving what Chatterton (2010) has called 'the urban impossible', or that which currently seems out of reach.

DOI: 10.4324/9780429275562-15

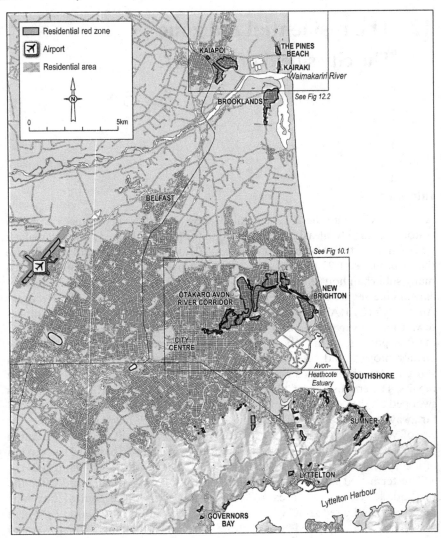

Figure 12.1 Location of the residential red zones in and around Christchurch.

Source: Data from http://koordinates.com/layer/3669/

It concludes by outlining some of the nascent characteristics of what a future Christchurch could look like, drawing on post-earthquake experimentation over the last decade.

Urban futures and the twenty-first-century city

The international debate over urban futures reflects a divide between, on the one hand, concern and fear that nothing much will change and, on the other, a belief that the city is always a work in progress open to being steered in progressive directions. The former position, which has often been character-ised by a long lament about neoliberalism, has its roots in unease about the rational designs of social order that characterise high modernism. James Scott identified an underlying drive by states to plan for legibility, so as to rationalise and standardise assets and resources into 'a more convenient for-mat', the convenience being of those who wield authority and power (Scott 1998, 3). Broadly, critics of such approaches instead promote values like citi-zen engagement and local knowledge. Theirs is a position that cities could be 'substantially distinct from the present and that organisations and individu-als have the power to shape the present to result in a different future' (Jeffrey and Dyson 2021, 654). This sort of vision parallels Boston's (2017) argument for anticipatory forms of governance as more effective means of managing emergent long-term issues, like the biodiversity and climate crises that are themselves the products of high modernism.

The more critical assessments of the state's post-earthquake planning in the city tend to focus on CERA's city centre blueprint. They do not see much progressive intent in this, even though it encompassed, indirectly, much that emerged from the Share an Idea engagement process. Rather the outcome is reported as having been imposed and experienced as such (Boswell 2021). A recurrent complaint is that bureaucracies such as CERA acted in post-political ways to evade difficult issues by removing them from the public arena: a direct reference to the implementation of the 100-day blueprint (Amore et al. 2017). The rationale for doing so echoes analyses of neoliberal risk manage-ment that furthers commercial interests (Amin 2013). Hence a focus on 'stability', or showing the world that the rebuilt city 'was open for business' (Dann 2021, 3). This was exactly CERA's intention, as Chapter 6 has shown. But the price, it is alleged, is a 'textbook case in how not to build back better', as those with resources and power have the least need to change (Matthewman and Byrd 2020). Thus the critique that a depoliticised process has narrowed the possibilities for transformation in the interests of predictability (Cretney 2019) and risk aversion (Lesniak 2016). Even those who accept the blueprint at face value, as a framework for regeneration, still see a need for 'significant community engagement and coherent governance' (Brand et al. 2020, 89).

In this respect, debates about the right to the city and the right to shape the city (Marcuse 2009) discuss a spectrum of ways of re-politicising citizen agency. At the broadest level, this involves using opportunities such as those wrought by disaster to 'fracture open' rather than foreclose political possibil-ities at a range of scales (Cretney 2019). An example is the propositional engagements of the transitional movement discussed in Chapter 8, which whilst not fully participatory nonetheless sought to offer 'provocations' for public trial. This approach accepts both failure and impermanence as poten-tial outcomes. A desire for improvisation does however depend upon facilita-tive responses from those in regulatory authority, to 'let go' in the face of uncertainty (Boswell 2021). Otherwise what Westbury (2015) characterised as the stifling impacts of bureaucratic processes can overshadow experimentation through seeming dedication to inventing rather than minimising difficulties. One successful example of how to ease the way in post-earthquake Christchurch has been Life In Vacant Spaces, a charitable trust discussed in Chapter 8, established with city council support. It mediates between owners of vacant properties and activators, whilst assisting the latter with health and safety requirements and insurance provision.

Such process-oriented approaches, instead of trying to build outcomes for an imagined future in terms of single rather than multiple meanings, embrace what might be termed the 'future-in-the-present'. Within the status quo, this necessarily involves more inclusive forms of decision-making in which there is an opportunity for a range of community voices to be heard (Amore et al. 2017). If this is to move beyond the post-political, however, it is also about wider ownership of decisions that shape the city, so that those who participate feel that they have a stake in the outcome, even if they do not necessarily agree with it. Ownership, and all that flows from that, such as motivation and community care of outcomes, springs from being recognised as a party to building the future in whatever form people in particular places determine that might take. This encompasses, for example, recognition of the interests of residents as well as of business, a process that has worked well in Christchurch when it has been tried. Examples include the valuing of children's voices in the design of the Margaret Mahy playground (Chapter 9), as well as a publicly focused strategy to produce Tūranga, the new central city library and information hub. Both of these projects also drew on the perspective and expertise of mana whenua, whose presence was all but silenced in the pre-earthquake city (Pickles 2016). In this respect, too, the future of the residential red zones is an opportunity to redraw the map of who has stakes in the city of the future.

Prefigurative forms of politics are a further development of engagement, being performative in the sense that people enact a vision of change through organisation, practices or design. The prefigurative is broadly progressive in seeking to 'embrace the potential of the future to be more inclusive, sustainable, and equitable than the present' (Jeffrey and Dyson 2021, 645). This involves the development of networks based on broadly shared commitments, trust and compromise, and the institutionalisation of practice to ensure durability. A good example internationally is the Transition Towns movement which models forms of urban life based on principles of mutuality, the sharing of experience and resources and low carbon futures (Aiken 2012). It is represented in the Canterbury region by Project Lyttelton (2022) in the port town, and by a number of rural and urban time banks. These are examples of how social practice can be 'thickened' through alliance building, coordination and persistence. The extent, however, to which the development of shared senses of purpose can be effective beyond the local in bringing real change remains open.

From the perspective of international debates about urban futures, Christchurch in the wake of the earthquakes was seen as a city full of possibility, indeed as 'a global city to watch' and 'a unique opportunity to rethink urban form' (Glaeser and Sassen 2011, 72). A decade on, however, critics consider that it has experienced a city centre and suburban rebuild, ringed by propertied greenfields, highlighted in Chapter 5, that embody and reinforce the attributes of twentieth-century modernist urbanism. The rebuild in this reading is seen as the product of prescriptive styles of planning that – with exceptions – tend to narrow rather than broaden the potential to deal with emergent social, economic and environmental crises (Pawson 2022). In contrast, if we think of the city 'not just as a static noun, but as an active verb' (Chatterton 2010, 235), then different futures might become imaginable. In view of the problems that cities planned in or of the twentieth-century lock in place, such as car and oil dependence, different futures are also very necessary. The question which therefore now arises is whether there is still opportunity to engage with the future-in-the-present within the residential red zone. And, in the spirit of Chapter 11, to consider how this area might be used to explore ways of living alongside those more-than-human actants which the Canterbury earthquake sequence brought vividly into human view.

Regeneration planning

If engaged citizens and critics disliked the process and direction of CERA's post-earthquake planning, nonetheless it is the case that the agency was never intended to have more than a five-year term. By 2016 central government accepted that a new direction was required, as outlined at the close of Chapter 4. The framework for the blueprint was then in place and the residential red zones had largely been cleared. The situation was ripe for an alternative approach, and beyond the city, in the town of Kaiapoi in Waimakariri District, this was already underway. This section therefore uses the Kaiapoi experience as a point of contrast with the Christchurch red zone (Figure 12.1). The latter, the Ōtakaro Avon river corridor, winds eastwards from the central city towards the Pacific Ocean. It covers 602 hectares, which subsided by up to 1.5 metres during the earthquakes (Hughes et al. 2015; Measures et al. 2011). With the process of red zoning, described in Chapter 10, it became probably the largest area of managed retreat in an urban setting in the world. Surely here was an opportunity to rethink urban form and process (Gundermann 2014).

The initial Crown offer to Ōtākaro Avon river corridor households was made in August 2011 (Chapter 10), and to those in the Kaiapoi red zones a few weeks later. By mid-2015, about 7000 houses had been removed in the river corridor, in addition to about a thousand in eastern parts of Kaiapoi and its small coastal settlements. It took a further year to demolish the 500 red-zoned houses in the Port Hills. All up, the cost to the Crown of these buyouts was about $1.5 billion, less than a quarter of which was recouped through insurance recoveries, mostly from the Earthquake Commission for land damage (Truebridge 2017). The process was managed by CERA. Broken footpaths and road surfaces, street lights and underground services were left in place. So too were many trees and shrubs, once community pressure forced a retention agreement in 2013 (AvON 2022). Former house sites were grassed, and whole blocks fenced to discourage fly-tipping whilst remaining accessible for active recreation. In 2016, Land Information New Zealand assumed responsibility for maintenance, although this involved little more than regular mowing and some pest control, as outlined in Chapter 11. The subsequent regeneration plan later described how 'previously vibrant residential neighbourhoods were deconstructed and replaced by open parkland' (Regenerate Christchurch 2019, 16).

This statement obscures the trauma of residents having to vacate their homes and neighbourhoods, as outlined in Chapters 3 and 10. One of the ways in which this trauma found an outlet was through community mobilisation to encourage future uses of the land that would respect red zone memories. This reflected a city-wide process of a 'post-quake proliferation of geography-based community groups' (Vallance and Carlton 2015, 33). Three such groups merged to establish the Avon Ōtākaro Network (AvON) in 2011 to urge that 'the lands be covenanted to remain in public ownership in perpetuity'. This was to guard against fears of land remediation, re-use and privatisation. Asking 'What is the future of the red zone?', one of the groups – the Campaign for a Memorial Reserve Covenant – devised maps showing either a 'Grey Zone of Uncertainty', which 'could blight the recovery of the east for years'; or a 'Green Zone' that 'provides the backbone and framework for the recovery of the east and a fitting memorial for those who lost their lives and their homes' (Scoop 2011b). Subsequently, AvON's (2022) vision also stated support for 'mana whenua aspirations for restored mahinga kai values'.

Other groups with broadly consistent values include Greening the Red Zone and the Avon-Ōtākaro Forest Park. But like those trying to engage actively in the future of the central city, all faced ongoing issues in trying to bring any sort of vision into reality. The city council, with a host of other post-earthquake problems to deal with, lacked the capacity to intervene in its red zones (Vallance 2015), which were anyway now owned by the Crown. CERA focused on land clearance rather than re-imagination. The situation that Amore et al. (2017) identified in the central city of 'non-decision making' with respect to community project activations was if anything amplified in the river corridor. Over a period of some years, AvON sought to coordinate a great range of community and some commercial project proposals for future uses, such as a heritage garden park, trails, community gardens, nature playgrounds and places of tranquillity. No mechanism was however available with which to institutionalise these, which is one of the conditions for the sustainability of prefigurative politics (Jeffrey and Dyson 2021). Projects with power and influence behind them, such as a local iteration of the Eden Project, backed by Eden International, and a rowing lake backed by a trust with private school and ruling political party affiliations, gained higher profiles than others, but ultimately no more traction.

This contrasted with the situation in Kaiapoi, where the district council continued to employ a distinctive pre-earthquake approach to community engagement that became known as 'the Waimakariri Way' (Vallance et al.

2019, 52). Although Waimakariri District covers more than 2000 square kilometres, it served a population of less than 50 000 at the time, concentrated in the towns of Rangiora and Kaiapoi (Figure 5.1). These conditions supported development of a civic value system based on trust and transparency. Drawing from experiences elsewhere, it acknowledged that communities have 'a vital role to play in disaster recovery' (Vallance 2015, 1287), as well as in finding workable solutions to problems and securing public legitimacy. An independent report commissioned by the council in August 2012 identified what mattered to residents in 'an integrated, community-based recovery framework' (Vallance 2013). Central to this framework was a hub in Kaiapoi, co-locating council teams and some central government agencies alongside community groups. This facilitated information flow and triangulation of data, proving invaluable when large-scale recovery programmes, such as the Kaiapoi Town Centre Rebuild, began (Vallance 2015). It also provided a foundation on which to build in subsequent engagement about Kaiapoi's red zone lands.

This engagement included the Canvas process in 2014, undertaken at the direction of the Minister for Earthquake Recovery, Mr Brownlee, and led by CERA with support from the Waimakariri council and Te Rūnanga ō Ngāi Tahu. Almost 600 people participated over a six week period, giving a clear message that they wanted 'a natural land- and water-based environment that reflects the Kaiapoi community and earthquake remembrance, and is inclusive of all residents, while enabling active and passive recreation' (CERA 2014, 10). The council was then tasked by the Minister with developing more detailed plans for the red zones in Kaiapoi and the coastal settlements of The Pines Beach and Kairaki (Figure 12.2). This was done – consistent with 'the Waimakariri Way' – through street corner and town meetings, expert review and a public hearing on submissions about the draft plan. An imaginative step was the use of a 3D model to enable public visualisation of the areas. The plan was readily approved by Minister Brownlee, perhaps because in addition to 41 hectares of greenspace, it included nine for mixed business use and 30 for light rural uses, such as grazing and cropping (DPMC 2016). It won the supreme accolade at both the New Zealand Planning Institute Awards and the Local Government Management Awards in 2017, reflecting the quality of the engagement process (Waimakariri District Council 2017).

Waimakariri therefore had an approved red zone plan before the start of any officially sanctioned consultation on the future of the Ōtākaro Avon corridor. This is partly accounted for by the difference in scale as the corridor is six times the size of the Kaiapoi red zones. It is also four times bigger than the city's central open space, Hagley Park, and twice the size of Manhattan's Central Park. It did not become an active focus of attention until Regenerate Christchurch was created under the Greater Christchurch Regeneration Act 2016, and then as a joint Crown-council agency. Regenerate's brief was to produce a spatial plan, working collaboratively whilst engaging and advocating 'effectively with communities and stakeholders' (Regenerate Christchurch 2021). There was however one special case inserted in the letter of instruction to the new agency by Minister Brownlee and the Mayor, Lianne Dalziel.

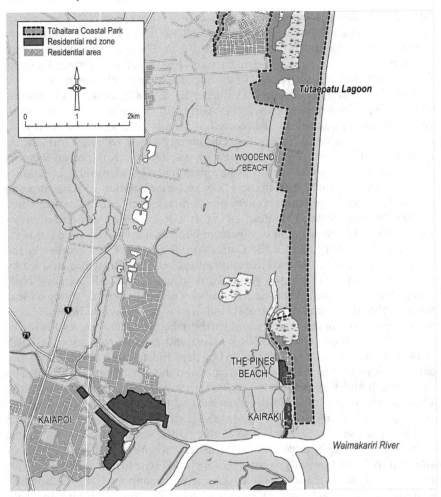

Figure 12.2 Kaiapoi red zones and the Tūhaitara Coastal Park.

Source: Adapted from Figure 12.1 and Pawson et al. (2019, 9).

This required that the East Lake Trust's proposal for a water sports facility of international standard be investigated (Brownlee and Dalziel 2016), the project having been a favourite of the Minister and his networks for some time (Stylianou 2016).

The design and engagement process that was implemented from late 2016 followed interwoven phases of technical analysis, community needs assessments, advisory group review and public exhibitions. Visioning and research events were used to develop an overarching vision, values and objectives. A design challenge was held, in which teams of people, half of them aged under 25, developed ten design options. These were put out for public consultation and shortlisted. Both the options and the shortlist were subject to the process of integrated assessment by a team of 50 people, drawn from a cross-section of community interests (Orchard 2017). The public exhibition of the final draft plan, held in the city centre over a three week period in late 2018, generated considerable interest, not least as it seemed to promise something. But a spatial design process offered nothing to project proponents other than a framework into which they might eventually fit, thereby prolonging the years of non-decision making. One proponent, a white water sports trust, shifted its interest to the Kaiapoi red zone. The East Lake Trust was also disappointed. Its proposal was deemed to be too environmentally risky, too likely to foreclose other options, and too expensive: despite this group using its intellectual capital to appeal the decision, unsuccessfully, to the Ombudsman (2018).

The Ōtākaro Avon River Regeneration Plan was finally approved by the Labour-led government's Minister for Greater Christchurch Regeneration in 2019. The chair of Regenerate Christchurch wrote in the foreword that the Ōtākaro Avon corridor 'represents a transformational opportunity of local, regional and national significance' (Regenerate Christchurch 2019, 6), balancing 'the needs and aspirations of current and future generations, iwi, and the wider public and private sectors'. Its seven objectives reflect the four well-beings. Both cultural and educational values are highlighted, for example, in the intent to provide 'a restored native habitat with good quality water so there is an abundant source of mahinga kai, birdlife and native species' and with the desire to establish 'a world-leading living laboratory' for experimentation and research. The plan sets out a continuous 345-hectare publicly accessible 'green spine' along the river from the city centre to the estuary. This will provide reafforested areas of native bush and wetland, the latter acting as retention basins as part of a city-wide flood management strategy. The spine is supplemented by three 'reaches' of additional land, that could accommodate project proposals and community spaces (Figure 12.3). If this map is performative, at least it leaves space for progressive engagements.

Experiments in governance

Such space was long to prove elusive. Nine years elapsed between the start of the earthquakes and the approval of the Ōtākaro Avon River Regeneration Plan; the Kaiapoi plan took six. Even then, there was no obvious way forward for those wishing to take advantage of this 'transformational opportunity'. The production of both plans had illustrated how public engagement in decision-making for recovery can work (Vallance 2015). In the case of Kaiapoi, this drew on prior post-earthquake experience encapsulated in 'the Waimakariri Way'. In the Ōtākaro Avon river corridor, it was developed from scratch by the new regeneration agency. Consultation and engagement, however, is one thing; action and implementation is another. In both cases, substantive change on the ground has been slow to materialise. This reflects uncertainties over responsibilities, land ownership and finance. As the chair

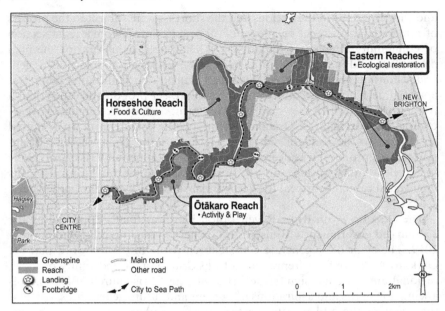

Figure 12.3 The Ōtākaro Avon River Corridor greenprint.

Source: Adapted from Regenerate Christchurch (2019, 48–9).

of Regenerate Christchurch wrote in the foreword to the draft Ōtākaro Avon regeneration plan, governance and accountability matters still needed to be decided, along with ownership of the land and funding commitments (in Regenerate Christchurch 2018b, 6). Significantly, this drew attention to the crucial issue of governance far more directly than the equivalent and anodyne text in the final Ministerially approved version (Regenerate Christchurch 2019).

Governance was more readily resolved – at least on the surface – in Kaiapoi. The ministerial foreword to the Waimakariri Residential Red Zone Recovery Plan stated that the Crown agreed to divest the land with this, in his words, 'underlin[ing] the Crown's ongoing commitment to a strong and resilient greater Christchurch' (DPMC 2016, 2). But there was no commitment to funding the implementation of the plan beyond this. There was however an agreement that the bulk of the land would be vested in the Waimakariri District Council, with five hectares of The Pines Beach and Kairaki coastal regeneration areas going to Te Kōhaka o Tūhaitara Trust. This Trust had been established under a 1998 Act of Parliament to manage an 800-hectare coastal park based on ecological, conservation and cultural values, and to uphold the mana of Ngāi Tahu by protecting and enhancing the environmental attributes of Tūtaepatu Lagoon (Figure 12.2). The Trust board comprises six trustees, three appointed by the Waimakariri council and three by Te Rūnanga o Ngāi Tahu (Pawson et al. 2019). As a working example of co-governance (Auditor-General 2016), the prior existence of the Trust enabled the divestment process to be completed in accordance with the Crown's treaty obligations under the Ngāi Tahu Claims Settlement Act 1998.

The situation was more complex in the Ōtākaro Avon corridor, not least because the partners in Regenerate Christchurch, the Crown and the city council, required that the plan omit specific recommendations about land ownership, future finance or governance. In effect this moved the critical implementation phase into a post-political cul-de-sac, hence the chair's statement about accountability noted above. The omission was not because work on these essential themes had not been done, although the need for agreement on them was noted on only one page of an 80-page document. The draft plan also observed that 'Direction about how the governance arrangements will incorporate the Treaty relationship will be provided by the Crown and Council' (Regenerate Christchurch 2018b, 63). Direction was however complicated by the shift in relationship between those two entities as a result of a change in central government in late 2017. The Labour-led administration that succeeded National's centre-right coalition that had been in power since 2008 made various electoral promises to post-earthquake Christchurch. These included a $300 million 'capital acceleration facility' to be spent on recovery as the city council determined, but within which were monies for the red zones.

The new government also signalled its desire to move to locally led regeneration in the city, the negotiation of which took time. The 'Global Settlement Agreement was signed in September 2019 between the Crown and Christchurch City Council to 'assist in the transition away from an extraordinary Crown presence in Christchurch and towards a normalised relationship with the Council' (DPMC 2019, 1). The two parties agreed on final cost shares and ownership of the anchor projects in the central city, as well as Crown divestment of red zone lands to the council. Responsibility for the management of those lands passed to the city council on 1 July 2020, and soon after the CERA-era fences began to come down. Divestment however was delayed for over two years more whilst Land Information New Zealand worked to amalgamate the mass of individual land titles and easements into tranches for transfer. A consultative group, Te Tira Kāhikuhiku, was established as an interim measure to advise the council and Land Information New Zealand on short-term transitional developments in the city's red zones. 'Short-term' meant for five years or less, which was of little use to those promoting longer-term initiatives, or requiring some security that their efforts would not be in vain.

The Global Settlement Agreement (2019, 7) recognised the role of Ngāi Tahu as Treaty partner, expecting that the city council would 'include Ngāi Tahu representation alongside other community representatives in … longer-term governance arrangements'. In practice, such representation is with Ngāi Tūāhuriri, the Ngāi Tahu hapū, or sub-tribe, that exercises cultural and customary authority in the river corridor. In the post-political vacuum that prevailed in the aftermath of the draft plan's appearance, a university-sponsored symposium for parties from interested agencies and groups had been held in May 2019 to discuss how such arrangements might be shaped. A follow-up analysis concluded that 'co-governance with mana whenua is a proven and essential model in the post-Treaty settlement era' (Pawson et al. 2019, 4). In Kaiapoi, this had been resolved by making use of the prior existence of the Tūhaitara Trust, although this did not provide for co-governance of the whole Kaiapoi red zone lands. Tūhaitara is however an example of a trust-based model, in which governance experience is shared between community and mana whenua. The long history of community engagement in the Ōtākaro Avon corridor had also generated anticipation of community involvement.

It was more than two years before governance discussions there were furthered when, under some community pressure, the mayor and upoko (head) of Ngāi Tūāhuriri sought advice from the previous government's Minister of Treaty Negotiations, Christopher Finlayson, who had nine years' experience of negotiating co-governance agreements. Co-governance extends greater participation in environmental decision-making to Māori, in recognition of their status as treaty partners (Finlayson and Christmas 2021). The advice proffered a spectrum of options drawing on cases already in operation. At the strong end of this spectrum are those that use the concept of legal personality (Morris and Ruru 2010) to recognise indigenous people's relationships with natural features and resources, notably Te Urewera and the Whanganui River in the North Island (Sanders 2018). Another example is the Waikato River Authority that exercises significant regulatory power over that river. At the weaker end of the spectrum are advisory bodies such as the Manawatu River Advisory Board, whose role is to offer advice to local government. The recommendation was for an entity in the middle of the spectrum, established as an independent board and likely under an Act of Parliament (Finlayson 2021).

In December 2021, the Christchurch City Council confirmed the intent to establish such a co-governance entity, followed in April 2022 by an agreement to set up a co-governance establishment committee to provide advice to the council and Ngāi Tūāhuriri 'on the development of the enduring co-governance entity/framework' for the river corridor (Christchurch City Council 2022b). Such an entity depends on building the relationships necessary for its effective function, a process that takes time, commitment and development of mutual trust (Auditor-General 2016). It will also require resourcing. The council's Long Term Plan for the period 2021–31 relied on an allocation of $40 million from the capital acceleration facility to start planting of the green spine, along with half that again from the Christchurch Earthquake Appeal Trust to commence recreational facilities such as pedestrian bridges and landings, initiated by the Parks Department with negligible community input. The Annual Plan for 2022–23 signalled council's commitment to delivery of the river corridor as an intergenerational project with an additional $20 million, as well as a river corridor activity plan intended to combine a range of council workstreams into a co-ordinated whole (Christchurch City Council 2022c).

How these council workstreams engage with the co-governance entity and the community remains to be seen. If a 'key dimension of prefigurative politics has been its commitment to action' (Jeffrey and Dyson 2021, 346), conditions to date have not been auspicious. Land Information New Zealand – still the landholder for the Crown – has required an arduous process of short-term licencing for any community initiative, and Te Tira Kāhikuhiku has had little money to assist these. Life in Vacant Spaces has facilitated some projects in a small block of the Burwood red zone. The Avon Ōtākaro Forest Park has managed to foster and extend modest areas of native planting with official permission. For comparison, in the Kaiapoi red zones, there is also little evidence of prefigurative actions on the ground, and the community enthusiasm of 'the Waimakariri Way' has been superseded by standard council park management initiatives. If the Ōtākaro Avon river corridor however is to be an engaging intergenerational project, means of facilitating the 'future in the present' remain to be negotiated. Regenerate Christchurch, which was disbanded in 2020, was therefore prescient in its concern about 'who will govern and have accountability for realising' the plan.

Christchurch futures

What then can be said about the potential of the river corridor as the city's 'field of dreams'? Whilst continuous delay has stalled projects and de-energised all but the most committed of activists, steps taken to date point to some potentially progressive outcomes. Four are considered here in terms of the corridor as a 50 or 100-year legacy project. The first is the commitment to co-governance, and what this may mean in terms of restoring mana whenua, or cultural and customary authority, to Ngāi Tūāhuriri. Their loss of access to the land and water resources of the corridor which occurred with colonisation and suburbanisation in the mid- to late-nineteenth century puts the last decade of inaction on the ground into perspective. During that time, the hapū's Matapopore Trust has completed a Cultural Design Strategy for the river. It observes that the areas of significance to Ngāi Tūāhuriri are in many cases those most affected by the 2011 earthquakes. Due to their 'low-lying, watery nature', these were rich in mahinga kai. This term 'quite literally means to work (mahi) the food (ngā kai), and refers to the seasonal migration of people to key food gathering areas to gather and prepare food to sustain them through the year' (Matapopore Trust 2017, 17–18).

One opportunity to be opened up by red zoning is therefore active restoration of indigenous values in the landscape. The Trust was established to provide cultural advice on Ngāi Tūāhuriri/Ngāi Tahu values, narratives and aspirations, and ways of weaving these into post-earthquake recovery projects. An example is the map (Figure 12.4) which highlights indigenous names and features in the Ōtākaro Avon corridor. The significance of water is clear: it is 'wāhi taonga, a treasured gift and is the life force of Papatūānuku' (Matapopore 2017, 17). Water provided the main source of sustenance and play, and determined the location of kāika (settlements), as well as being a

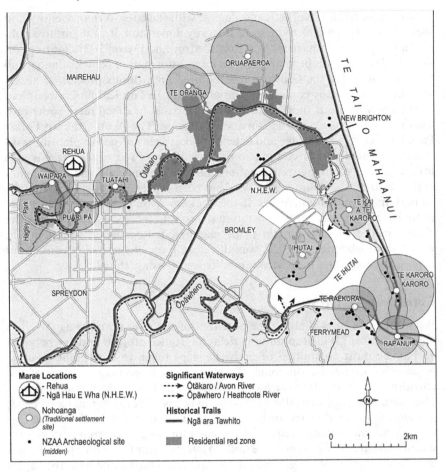

Figure 12.4 Features of significance to mana whenua.

Source: Adapted from Regenerate Christchurch (2019, 15).

medium of travel. It is the basis of the places, philosophies and practices of mahinga kai. This is more than a matter of historical memory. Section 6 of the Te Rūnanga o Ngāi Tahu Act 1996 mandates the rūnanga (tribal council) to protect the beneficial interests of all members of Ngāi Tahu as the Treaty partner. In terms of post-earthquake legislation, this includes an active role in the governance of recovery (Chapter 4). Whilst co-governance in this context has generated little community comment, or indeed debate at council level, the implications of adoption of such arrangements in broader spheres of government policy is now drawing wider public scrutiny (eg. Salmond 2021).

A second potentially progressive outcome from the former residential red zone is the provision of a more 'restorative city'. Although this concept is commonly used to describe urban design for mental health and well-being (Roe and McCay 2021), it can be extended in this context to include cultural well-being. The adoption of co-governance, above, is a partial recognition of the reclamation 'of the right of Indigenous peoples to once again govern themselves in their own lands' (Jackson 2021, 135). The plan for ecological restoration, encapsulated in the greenprint (Figure 12.2), likewise respects the red zone memories of former residents whose concerns were central to the foundation of the Avon Ōtākaro Network. In both contexts, 'restorative' implies respect for cultural and social belonging and inclusion. It also captures the central importance to prosperity and well-being of human health and the vitality of the more-than-human world on which people depend. This is explicitly recognised in the three pillars of the Avon Ōtākaro Forest Park concept, which pictures the interdependence of the corridor's blue and green ecosystems with local communities. Features like clean air and water, biodiversity and restored habitat provide the means for the delivery of health benefits and community engagement as well as opportunities for active and passive recreation (Avon Ōtākaro Forest Park 2022).

Restorative benefits can be delivered through use values but are likely to be multiplied if citizens and community groups can participate actively in their delivery and management. Just as co-governance is an active alternative to the standard model of local government authority, so too are means that open improvisational opportunities for citizen action. One sphere in which this is becoming widely practised in urban and rural New Zealand is community management of predator control to foster survival of native fauna (Burge et al. 2021). This will also be required throughout the river corridor to restore biodiversity. A potential flagship project in this respect, introduced in Chapter 11, is a long-delayed proposal to use part of the Burwood red zone and adjacent existing reserve of Travis wetland as a fenced eco-sanctuary as a safe haven in which native species can breed, and from which they can disperse more widely in the city (Waitākiri Eco-sanctuary 2022). Such initiatives are consistent with the concept of 'anticipatory innovation governance', championed by the OECD, which describes continuous local adaptation and experimentation as means of addressing complex place-based problems as they emerge. This applies especially to those 'issues that are too complex or evolve too quickly for orthodox policy responses' (OECD 2021).

This points to a third area of progressive outcomes, that of climate change response. The Waimakariri Way and the comprehensive process of engagement undertaken by Regenerate Christchurch have already provided some invaluable civic experience in collaborative approaches towards resolving wicked problems. The policy of red zoning was developed in the first place as these earthquake-damaged lands are at heightened risk of climate-related inundation in coming decades. The process of community withdrawal has long been all but complete, as have formal processes of regeneration planning for the spaces opened up by retreat. Despite the time that has since elapsed, the opportunity now exists to build on the collaborative approaches that characterised the planning process to involve community partners in the implementation of the plans (Pawson 2022). When the politics of neoliberalism has widened inequalities and reduced trust in government internationally, there is a growing need for such deliberative governance processes that focus as much on process as on outcomes. These are well-suited to intricate situations characterised by value-driven dilemmas and long-term time horizons (OECD 2020).

Adaptation is inevitably place-specific and can only be shaped and implemented through local collaborations. When in 2015, the Christchurch City Council published technically sourced maps identifying thousands of properties at risk of coastal erosion or flooding, there was an outcry given the lack of local input (Peat 2018). Yet about the same time, the Parliamentary Commissioner for the Environment (2015) recommended that councils be provided with guidance on engaging with coastal communities and avoiding surprises. This is echoed by the Intergovernmental Panel on Climate Change Sixth Assessment Report on Impacts, Adaptation and Vulnerability (IPCC 2022). One of its drafting authors describes the prospects for 'climate-resilient development' for the 11 per cent of the world's population that live on low elevation coasts as 'dismal', with a 'coastal adaptation gap' that needs to be closed quickly. Again, among the key responses necessary are governance capabilities and facilitative participation (Glavovic 2022). The red-zoning process was hardly participatory but exists as a large-scale response from which other places can learn, and against which other vulnerable parts of the Christchurch coast might develop alternatives.

There is an element of experimentation about such responses, and this is the fourth area of potentially progressive outcomes. Co-governance itself, being place-specific and based on relationship building, is necessarily experimental. The co-governance entity will have to provide a secure and facilitative framework for even modest prefigurative initiatives to succeed, be these community gardens, art installations or culture trails. There is potential however to host more ambitious technical experimentation, in line with the objective of the plan to establish an Ōtākaro Living Laboratory (Chapter 11). An example is small-scale structural prototypes, for housing or recreational activity, designed to cope with rising waters (University of Washington 2020). A broader approach would ask what part could the red zone lands can play in research and development, as well as pilot and demonstration projects for climate security? Furthermore, how might an ongoing process of red zoning over time bring into being new types of urban form: in which progressively abandoned areas are transformed into green buffers to play protective and restorative roles in the city of the future (Gundermann 2014)?

If such ideas are to be shepherded, and citizen agency and prefigurative action is not to be frustrated, as it has been for much of the last decade, by a lack of support and facilitative governance, then boldness of vision will be required. This will not occur if the co-governance structure in prospect replicates the time-honoured bureaucratic model of contracting internally for local government delivery. In that case, the right to shape the city and the opportunity to experience it 'as an active verb' will be lost. Rather than risk mitigation, the red zones provide an opportunity for risk-taking, so that the enthusiasm of those who desire to engage with, contribute to and build new futures is supported, encouraged and valued. This is the essence of 'anticipatory innovation governance', without which any ambition to meet climate challenges, to build 'national park cities' or to use co-governance structures in moving towards sustainable lifestyles will be compromised.

Conclusion

One of the insights of comparative studies of earthquake recovery is it always takes time (Johnson and Olshansky 2017). This applies to conventional styles of centrally directed rebuilding, such as the anchor projects of the central city (Figure 4.4), as well as the more open-ended processes of planning in the former residential red zones. In Waimakariri District, the Kaiapoi red zone recovery plan built on a solid record of community engagement and was approved relatively soon in the recovery process, in 2016. That was the year that Regenerate Christchurch was set up to initiate a process of engagement that three years later produced an approved Ōtākaro Avon regeneration plan. Even then, slow progress towards securing long-term governance has forestalled the prospect of the corridor as 'a field of dreams' and work on the ground has been slow. That however is only one perspective. In contrast, for Ngāi Tahu and Ngāi Tūāhuriri, the opportunity to create an effective model of co-governance, in recognition of their own memories in the landscape, is one that has been more than 150 years in the waiting.

Whether or not the red zones will become experiments in the twenty-first-century urbanism and adaptation to changing sea levels is therefore still an open question. Progress to date has been largely procedural, rather than substantive (Vallance 2015), but perhaps this is not an achievement to underestimate. Regenerate Christchurch was established to engage, advocate, work collaboratively and 'provide independent advice to decision makers' (Regenerate Christchurch 2021), in other words, to develop and practice forms of network governance after the state-led years of CERA. Potentially this – along with the 'Waimakariri Way' – has built capability and social capital that can be transferred to other contexts, to construct for example locally-rooted restorative, biodiversity and climate change strategies. With good governance, these can enable evolving relationships between people and place, and between the human and the more-than-human, as explored in the previous chapter. So experience in the red zones has already demonstrated that there are progressive ways of doing the urban, although it has also shown that this certainly takes time. Perhaps in time, it may also reclaim some of the international attention that the transitional movement garnered for the city in the wake of the quakes.

13 Conclusion

In this book, we have sought – as indicated in Chapter 1 – to 'provide a framework in which to assemble the multiple experiences and realities' of post-earthquake recovery in Christchurch. This concluding statement seeks to reiterate and summarise key aspects of that framework, using four broad themes. This is done in the spirit, also identified in Chapter 1, of distrusting 'singular meta-narratives and headline conclusions'. Accordingly, it is less a summary and more an identification of themes that have surprised and intrigued us in the course of our collaboration. These themes focus first, following Donna Haraway (2016), on 'staying with the trouble'; on second, Alain Badiou's (2005) contention that 'events' produce change based on both familiar outcomes and alternative potentials; on third, that post-disaster experiences are often quite divergent for different places and groups of people; and on fourth, the extent to which the city is now better prepared for what May Joseph (2013b, 3) has called an 'ecological future' in the twenty-first century.

The first thing this book has attempted to do is to pay attention to the human experience of post-earthquake recovery, rather than to identify too closely with the tidiness and optimism of planners and architects inherent in such artefacts as the 100-day blueprint. In this sense, we have adapted Donna Haraway's call (2016) to 'think-with, live-with and be-with' a city and citizenry still coming to terms with the rupture of earthquakes. This is a rupture that has been deeply felt experientially, economically, socially and culturally. As an event, the earthquake sequence has, as Matthew Galloway (2021, 276) puts it, 'exposed the façade of colonial narratives that have informed much of the city's history and identity'. The fate of the Anglican cathedral, which still lies in ruins although now under long-term repair, symbolises the perils of easy adoption of any post-earthquake narrative of progress. Yet it was the stylised cathedral spire, soaring into a blue sky above a carpet of green, that for many years before 2010 was used as a logo to symbolise Christchurch, 'the garden city'. The earthquakes therefore unleashed not only material destruction but also destruction of the city's idea of itself (Pickles 2016).

In this sense, to stay with the trouble is to be open to new possibilities, and to recognise and welcome into the present those human and more-than-human interests that were elided by the blue sky and green carpet. If Christchurch

DOI: 10.4324/9780429275562-16

was a proud colonial city, displaying in its nineteenth-century architecture the achievement of 'terraforming' and 'improvement' of the wetland 'swamps' on which it was planted, then the hazards of liquefaction and lateral spread have punctured any sense of stability. So too the experience of many businesses and householders trying to resolve insurance claims in a country where heavy levels of insurance had been presumed to provide confidence and security. In their place has come both uncertainty and opportunity. Halting pre-earthquake steps to engage with mana whenua have found far more impetus in the rebuild, and the swamp is now, through embracing blue-green spaces and corridors, experienced in more generative terms. It has proved hard however to manage the tensions between speed and deliberation in rebuilding. In a considered assessment of the experience of six countries, Johnson and Olshansky conclude that recovery from disasters is 'always complex, takes a very long time, and is never fast enough for affected residents' (2017, 329).

Our second theme concerns the nature of social change that follows the 'event' of a rupture such as the Canterbury earthquake sequence. Badiou (2005) characterises this in two overlapping ways: either through resort to familiar conduct managed in terms of dominant ideologies or through discovery of alternative actions that previously had been submerged or hidden. The 'overlap' is significant. The central city blueprint, for example, was shaped by the neoliberal state's desire for a quick and effective rebuild, designed to hold in place and attract business investment. Yet it drew on some of the key ideas that had emerged in the prior 'citizens' plan' for the centre, even if it did not continue to provide a place for deliberative engagement once approved. In this respect, the New Zealand government adopted 'one of the most visible governmental responses to disaster', with the creation by legislation of a new agency to act as a focal point for recovery and to protect its own interests (Johnson and Olshansky 2017, 314). Yet this too must be qualified. The Canterbury Earthquake Recovery Act of 2011 was time limited to five years, after which it was replaced by the Greater Christchurch Regeneration Act, which attempted to share greater responsibility between the Crown and local interests. In turn, the growing desire for this shift saw key aspects of the Regeneration Act itself rendered redundant before it had run its own five-year term.

Badiou's observation is also relevant in a deeper sense. Much of the Christchurch rebuild has been shaped by path dependency. There was never a serious question that the city centre would be rebuilt in the same place, despite the instability and hazards of the site. Yet at the same time, post-earthquake planning instruments have exacerbated a long-run tendency towards urban sprawl in terms of greenfield development, notably in the south-west of the city and in satellite towns in the wider urban region. The Crown has invested simultaneously in an expensive city centre blueprint whilst underwriting a new motorway system that fuels the greenfields. In these ways, post-earthquake Christchurch has often been criticised for reproducing the attributes of twentieth-century cities, whilst foregoing the opportunities to embrace new forms of twenty-first-century urbanism (Pawson 2022). It is seen as not delivering on the promise of the immediate post-quake years, when the transitional movement, driven informally by younger people in the arts and education (eg. Gap Filler, the Student Volunteer Army) as well as some in business (eg. the container mall) earned the city a reputation as a place to watch. Paradoxically, however, it may be that the city's new (but largely traditional) landscapes of consumption provide many of its inhabitants with exactly what they need for a return to familiar ways.

This claim intersects with our third theme: that post-earthquake experiences have been quite variable and at times divergent for different places and groups of people. Maybe in one sense Fiona Farrell (2015, 70) was correct when she wrote that 'An earthquake is the most egalitarian of disasters. It strikes an entire region in one blow'. After all, most of the city's brick chimneys were destroyed or damaged in the early hours of 4 September 2010 when the earthquake sequence began (Figure 2.3). But many other maps in this book show sharply different experiences. Housing was particularly badly affected in the eastern parts of the city due to the spatial incidence of liquefaction, and places in the east, in coastal areas and on the hills, were so impacted as to be red-zoned. The well-being of people with few resources to manage recovery has been particularly shaped by insurance battles that lasted for years for some households. In turn, this led to the creation of effective networks of resident activism, forcing a response from state institutions. Likewise, protest and mobilisation in the eastern suburbs have ensured investment in new schools and recreational facilities; in this respect, the city council has been more responsive to need on the ground than government in Wellington often proved to be. This illustrates one of Johnson and Olshansky's (2017, 326) key recommendations for recovery, that 'National governments are important sources of money, technical support, guidance and oversight, but local governments are best suited to implement recovery and devise actions appropriate to their needs'.

The conflict between central government direction and local needs and actions has been highlighted in another aspect of the Christchurch experience. The central city blueprint was in part a product of the Crown's concern that without a massive input of its own resources in the form of investment in anchor projects (albeit shared with the city council) and its ability to force land title amalgamations, then business investment would be lost to other centres and any international interest in commercial property redevelopment would evaporate. Yet the post-earthquake rebuild attracted little international interest anyway, with much of it being financed by local property investors turned developers drawing on insurance pay-outs and their own resources. Their motives rather than being shaped by the economically rational alone have reflected the more affective quality of emotional connection with their home town. Ironically, the blueprint's intention that the anchor projects would lead the rebuild, providing a framework within which business investment would be secured, also transpired to be in error. Rather local property investment has often come first, with many of the key anchor projects being delayed for years. Some have yet to be finished, leading to difficult resource allocation decisions a dozen or more years on from 2010–11.

Our final point concerns whether the post-earthquake city is now better placed to meet the challenges of the twenty-first century than it would otherwise have been. There is no easy answer to this, but if the city is a 'petri dish', in which 'the laboratory is humming, and the great experiment' is underway (Farrell 2015, 323), then what are the indications? It is easy to be critical. The path-dependent re-shaping of the city has baked in the very twentieth-century attributes of a suburbanised and auto-dependent suburban landscape that internationally have produced carbon-intensive ways of life. Some of the city's key anchor projects do too, being reliant on earlier expectations about levels of international tourism. Te Pae, the massive downtown convention centre, with its global pretentions, and Te Kaha, the as-yet unfinished 'multi-use arena', are perhaps the ultimate expressions of what might be called a pre-Covid, carbon agnostic mindset. At the same time, the blueprint has delivered a re-imagined city centre that is more amenable to both pedestrians and cyclists, and the whole urban area is now criss-crossed by cycle paths that provide the essential infrastructure for alternative mobilities, even if these have as yet made little difference to mode shares.

But perhaps the real gains in terms of readiness for a post-earthquake, climate-challenged future lie elsewhere. If the often criticised top-down approach to recovery gave little acknowledgement to the work of post-earthquake transitional organisations, the transitional movement nevertheless unleashed enthusiasm for collective and inclusive ways of working. Arguably the legacy of this has been the creation of new political spaces of hope and action that have been echoed in many arenas such as resident activism, new forms of voluntarism and care, and experiments in community engagement. We have identified many very real examples of what Rebecca Solnit (2009) called 'the extraordinary communities that arise in disaster'. Many of these initiatives have been supported by novel or experimental governance arrangements: new ideas and ways of being in the city are negotiated and grounded in changing forms of governance. The collaborative process that produced the pre-earthquake Urban Development Strategy, for example, provided some direction for its post-earthquake recovery. Experiments in co-governance elsewhere have informed a strategy for the delivery of future options for the residential red zone. And citizen experience of many forms of activism and deliberation may yet prove to be invaluable as the city navigates the emerging uncertainties of climate change and sea level rise from now on.

References

ACTIS (2022). Aranui Community Trust Incorporated Society. Retrieved from: https://actis.org.nz

Adams-Hutcheson, G. (2015). Voices from the margins of recovery: Relocated Cantabrians in Waikato, *Kōtuitui: New Zealand Journal of Social Sciences Online* 10 (2), 135–43.

Adger, W.N. (2000). Social and ecological resilience: Are they related? *Progress in Human Geography* 24 (3), 347–64.

Adger, W.N. (2006). Vulnerability, *Global Environmental Change* 16, 268–81.

Aiken, G. (2012). Community transitions to low carbon futures in the transition towns network (TTN), *Geography Compass* 6 (2), 89–99.

Allot, A. (2021). Suspected dog attack causes 500 critically endangered gulls to abandon nests, *Stuff*, 23 November. Retrieved from: www.stuff.co.nz/environment/127069075/suspected-dog-attack

Amin, A. (2013). Surviving the turbulent future, *Environment and Planning D: Society and Space* 31 (1), 140–56.

Amore, A., Hall, C.M. and Jenkins, J. (2017). They never said 'Come here and let's talk about it'. Exclusion and non-decision-making in the rebuild of Christchurch, New Zealand, *Local Economy* 32 (7), 617–39.

Andersen, C. and McLellan, W. (2014). An EPIC view of innovation, in B. Bennett, J. Dann, E. Johnson and R. Reynolds eds., *Once in a Lifetime. City-building after Disaster in Christchurch*. Freerange Press, Christchurch, 362–5.

Anderson, S., Bennett, M., Birdling, M. et al. (2010). Academics call for rethink over earthquake law, *Stuff*, 28 September. Retrieved from: www.stuff.co.nz/the-press/news/christchurch-earthquake-2011/canterbury-earthquake-2010/4174748/Academics-call-for-rethink-over-earthquake-law

Anderson, V. (2012). 'Ministry' created to 'water the seeds of awesome', *Stuff*, 15 June. Retrieved from: www.stuff.co.nz/the-press/news/7106627/Ministry-created-to-water-the-seeds-of-awesome

Ansell, C. and Gash, A. (2008). Collaborative governance in theory and practice, *Journal of Public Administration Research and Theory* 18 (1), 543–71.

Ansley, B. (2011). *Christchurch Heritage. A Celebration of Lost Buildings and Streetscapes*. Random House, Auckland.

Anthony, J. (2021). Battle of the convention centres, *Stuff*, August 1. Retrieved from: www.stuff.co.nz/business/industries/125861975/battle-of-the-convention-centres-three-new-venues-worth-14b-provide-onceinalifetime-opportunity

Ardagh, M. and Deely, J. (2018). *Rising from the Rubble. The Health System's Extraordinary Response to the Canterbury Earthquakes*. Canterbury University Press, Christchurch.

Arnouts, R., van der Zouwen, M. and Arts, B. (2012). Analysing governance modes and shifts – governance arrangements in Dutch nature policy, *Forest Policy and Economics* 16, 43–50.

Atkinson, B., Baxter, S., Ver Berkmoes, R., Bindloss, J. and Blasi, A. (2013). *Lonely Planet's Best in Travel 2013*. Lonely Planet Publications, Hawthorn.

Auditor-General (2016). *Principles for Effectively Co-governing Natural Resources*. Presented to the House of Representatives under section 20 of the Public Audit Act 2001.

AvON (2022). Our history of achievements. Avon Ōtākaro Network. Retrieved from: http://www.avonotakaronetwork.co.nz/about-us/history-of-achievements.html

Avon-Heathcote Estuary Ihutai Trust (2022). Our objectives. Retrieved from: www. estuary.org.nz/about/the-trust.html

Avon Otakaro Forest Park (2022). A forest park vision for the Christchurch red zone. Retrieved from: www.aofp.co.nz/our-plan

Badiou, A. (2005). *Being and the Event*. Continuum, London.

Baker, T. and McGuirk, P. (2021). Out from the shadows? Voluntary organisations and the assembled state, *Environment and Planning C, Politics and Space* 39 (7), 1338–55.

Ballard, S., Benson, T., Carter, R., Corballis, T., Joyce, Z., Moore, H., Priest, J. and Smith, V. (2015). *A Transitional Imaginary. Space, Network and Memory in Christchurch*. Harvest, Christchurch.

Barclay, C. (2021). Christchurch's $1 million suburb club expands as housing bonanza continues, *Otago Daily Times*, 3 April. Retrieved from: www.odt.co.nz/star-news/ star-lifestyle/star-home-and-gardening/christchurchs-1-million-suburb-club-expands-housing

Bassett, K. (2008). Thinking the event: Badiou's philosophy of the event and the example of the Paris Commune, *Environment and Planning D: Society and Space* 26 (5), 895–910.

Bayer, K. (2012). Christchurch schools confront confusion, *New Zealand Herald*, 14 September. Retrieved from: www.nzherald.co.nz/nz/christchurch-schools-con front-confusion/GJLJUXAELMF3SBHAZLKXTWWAXA/

Bayer, K. (2017). Aftershock: The debacle of shoddy Christchurch earthquake repairs, *New Zealand Herald*, 3 September. Retrieved from: www.nzherald.co.nz/nz/ aftershock-the-debacle-of-shoddy-christchurch-earthquake-repairs/ LEA5ARBI32NIQPB6MRSJHDF4OQ/

Beaglehole, B., Mulder, R.T., Boden, J.M. and Bell, C.J. (2019). A systematic review of the psychological impacts of the Canterbury earthquakes on mental health, *Australian and New Zealand Journal of Public Health* 43 (3), 274–80.

Beauregard, R.A. (2020). *Advanced Introduction to Planning Theory*. Edward Elgar, Cheltenham.

Bednar, D., Henstra, D. and McBean, G. (2019). The governance of climate change adaptation: Are networks to blame for the implementation deficit? *Journal of Environmental Policy and Planning* 21 (6), 702–17.

Begg, A., D'Aeth, L., Kenagy, E., Ambrose, C., Dong, H. and Schluter, P.J. (2021). Wellbeing recovery inequity following the 2010/2011 Canterbury earthquake sequence: Repeated cross-sectional studies, *Australian and New Zealand Journal of Public Health* 45 (2), 158–64.

Bennett, B. (2016). Interview with Barnaby Bennett, in M. Lesniak, Rebuilding/ Reimagining, Post-disaster Christchurch. Unpublished Masters Altevilles dissertation, University of Lyon, Lyon, 151–60.

Bennett, B., Boidi, E. and Boles, I. eds. (2012). *Christchurch. The Transitional City Pt IV.* Freerange Press, Christchurch.

Bennett, B., Dann, J., Johnson, E. and Reynolds, R. eds. (2014a). *Once in a Lifetime. City-building after Disaster in Christchurch.* Freerange Press, Christchurch.

Bennett, B., Dann, J., Johnson, E. and Reynolds, R. (2014b). Conclusion, in B. Bennett, J. Dann, E. Johnson and R. Reynolds eds., *Once in a Lifetime: City-building after Disaster in Christchurch.* Free Range Press, Christchurch, 478–80.

Bennett, J. (2010). Ontology, sensibility and action, *Contemporary Political Theory* 14 (1), 82–9.

Berke, P., Backhurst, M., Day, M., Ericksen, N., Laurian, L., Crawford, J. and Dixon, J. (2006). What makes plan implementation successful? An evaluation of local plans and implementation practices in New Zealand, *Environment and Planning B: Planning and Design* 33 (4), 581–600.

Beuys, J. and Harlon, V. (2004). *What is Art? Conversation with Joseph Beuys.* Clairview, Forest Row.

Blaikie, P., Cannon, T., Davis, I. and Wisner, B. (1994). *At Risk. Natural Hazards, People's Vulnerability, and Disasters.* Routledge, London.

Blair, M.J. and Mabee, W.E. (2020). Resilience, in A. Kobayashi ed., *International Encyclopedia of Human Geography*, 2nd edn., Elsevier, Amsterdam, 11, 451–56.

Blundell, S. (2006). Down at the mall, *New Zealand Geographic* 77, 37–53.

Blundell, S. (2021). A tale of two churches, *North & South*, February, 48–57.

Boffa Miskell Limited (2020). *RRZ Pest Animal Surveys 2020: Biosecurity Assessment Report.* Report prepared for Land Information New Zealand.

Bohan, E. (2022). *Heart of the City. The Story of Christchurch's Controversial Cathedral.* Quentin Wilson Publishing, Christchurch.

Borella, J., Quigley, M., Riley, M., Trutner, S., Jol, H., Borella, M., Hampton, S. and Gravley, D. (2020). Influence of anthropogenic landscape modifications and infrastructure on the geologic characteristics of liquefaction, *Anthropocene* 29, 100235.

Boshier, P. (2017). *Disclosure: An Investigation into the Ministry of Education's Engagement Processes for School Closures and Mergers*, New Zealand Parliament. Retrieved from: www.ombudsman.parliament.nz/sites/default/files/2019-03/ Disclosure.pdf

Boston, J. (1995). *The State under Contract.* Bridget Williams, Wellington.

Boston, J. (2017). *Safeguarding the Future. Governing in an Uncertain World.* Bridget Williams Books, Wellington.

Boston, J., St. John, S. and Dalziel, P. (1999). *Redesigning the Welfare State in New Zealand. Problems, Policies, Prospects.* Oxford University Press, Oxford.

Boswell, R. (2021). Play, politics and the production of space: DIY urbanism in post-earthquake Christchurch. Unpublished PhD thesis, University of Auckland, Auckland.

Bowring, J. (2021). Tree sense of place, in S. Goldsmith ed., *Tree Sense. Ways of Thinking about Trees.* Massey University Press, Auckland, 105–21.

Bowron, J. (2019). Christchurch squatters see the beauty of the city with a gap-toothed smile, *Stuff*, 18 July. Retrieved from: www.stuff.co.nz/the-press/news/ christchurch-earthquake-2011/117263021/christchurch-squatters-see-the-beauty-of-the-city-with-a-gap-toothed-smile

Brand, D., Allen, N. and O'Donnell, G. (2020). The New Zealand experience of a design-led approach to post-earthquake recovery in Christchurch, *Urban Studies and Public Administration* 3 (3), 89–115.

Brandes Gratz, R. (2015). *We're Still Here Ya Bastards: How the People of New Orleans Rebuilt Their City*. Nation Books, New York.

Breakthrough Facilitation (2019). Breakthrough Facilitation Facebook page. Retrieved from: www.facebook.com/breakthroughfacilitation/

Brenner, N. and Theodore, N. (2002). Cities and the geographies of 'actually existing neoliberalism', *Antipode* 34 (3), 349–79.

Brown, T. (2009). *Change by Design: How Design Thinking Transforms Organizations and Inspires Innovation*. HarperCollins, New York.

Brown, C., Seville, E. and Vargo, J. (2013). *The Role of Insurance in Organisational Recovery Following the 2010 and 2011 Canterbury Earthquakes*. Resilient Organisations, University of Canterbury, Christchurch.

Brownlee, G. (2016). Third reading, Greater Christchurch Regeneration Bill, *Hansard*, 31 March. Retrieved from: www.parliament.nz/en/pb/hansard-debates/rhr/document/51HansD_20160331_00000012/greater-christchurch-regeneration-bill-third-reading

Brownlee, G. and Dalziel, L. (2016). Letter of expectations for Regenerate Christchurch, 14 April. Retrieved from: www.regeneratechristchurch.nz/assets/Uploads/regenerate-christchurch-letter-of-expectations-14-april-2016.pdf

Buchanan, A.H., Carradine, D., Beattie, G. and Morris, H. (2011). Performance of houses during the Christchurch earthquake of 22 February 2011, *Bulletin of the New Zealand Society for Earthquake Engineering* 44 (4), 342–57.

Buchanan, A.H. and Newcombe, M.P. (2010). Performance of residential houses in the Darfield (Canterbury) earthquake, *Bulletin of the New Zealand Society for Earthquake Engineering* 43 (4), 387–92.

Bulkeley, H., Coenen, L., Frantzeskaki, N., Hartmann, C., Kronsell, A., Mai, L., Marvin, S., McCormick, K., van Steenbergen, F. and Palgan, Y.V. (2016). Urban living labs: Governing urban sustainability transitions, *Current Opinion in Environmental Sustainability* 22, 13–17.

Bullard, R.D. (2018). *Race, Place, and Environmental Justice after Hurricane Katrina. Struggles to Reclaim, Rebuild, and Revitalize New Orleans and the Gulf Coast*. Routledge, New York.

Buller, H. (2014). Where the wild things are: The evolving iconography of rural fauna, *Journal of Rural Studies* 20, 131–41.

Bullock, C. (2014). *Quake Cats. Heart-warming Stories of Christchurch Cats*. Random House, Auckland.

Burge, O.R., Innes, J.G., Fitzgerald, N. Guo, J., Etherington, T.R. and Richardson, S.J. (2021). Assessing the habitat and functional connectivity around fenced eco-sanctuaries in New Zealand, *Biological Conservation* 253, 108896.

Burn, I. (2020). Personal communication, 17 March. Ian Burn was General Manager, Delta Community Support Trust.

CAE (1997). *Risks and Realities. A Multi-disciplinary Approach to the Vulnerability of Lifelines to Natural Hazards. Report of the Christchurch Engineering Lifelines Group*. Centre for Advanced Engineering, University of Canterbury, Christchurch.

Campagnolo, S. (2015). Personal communication, 5 March. Sarah Campagnolo was then project co-ordinator with Greening the Rubble.

Campbell, K.T. (2014). The Shaken Suburbs. The Changing Sense of Home and Creating a New Home after Disaster. Unpublished Master of Science thesis, University of Canterbury, Christchurch.

CanCERN (2014). CanCERN weekly newsletter #131, 27 June. Retrieved from: https://quakestudies.canterbury.ac.nz/store/object/216073

Canterbury District Health Board (2021). *Evaluation of the All Right? Campaign COVID-19 Response - Getting Through Together*. Retrieved from: www.cph.co.nz/wp-content/uploads/GTTEvaluationReport.pdf

Carlton, S. (2013). Share an idea, spare a thought: Community consultation in Christchurch's time-bound post-earthquake rebuild, *Journal of Human Rights in the Commonwealth* 2, 4–13.

Carlton, S. and Mills, C.E. (2017). The student volunteer army. A 'repeat emergent' emergency response organisation, *Disasters* 41 (4), 764–87.

Carlton, S., Nissen, S. and Wong, J.H.K. (2021). A crisis volunteer 'sleeper cell'. An emergent, extending and expanding disaster response organisation, *Journal of Contingencies and Crisis Management*. https://doi.org/10.1111/1468-5973.12381

Carlton, S., Nissen, S., Wong, J.H.K. and Johnson, S. (2022). 'A shovel or a shopping cart': Lessons from ten years of disaster response by a student-led volunteer group, *Natural Hazards* 111, 33–50.

Carlton, S. and Vallance, S. (2013). *An Inventory of Community-led and Non-governmental Organisations and Initiatives in Post-earthquake Canterbury*. Lincoln University, Canterbury. Retrieved from: https://communityresearch.org.nz/wp-content/uploads/formidable/Final-Inventory.pdf

Carlton, S. and Vallance, S. (2017). The commons of the tragedy: Temporary use and social capital in Christchurch's earthquake-damaged city, *Social Forces* 96 (2), 831–50.

Central City Recovery Plan (2012). *Central City Recovery Plan. Te Mahere 'Maraka Ōtautahi'*. Canterbury Earthquake Recovery Authority, Christchurch.

CERA (2011). Crown offer to residential insured property owners in the Canterbury earthquake affected red zones. Paper 1. Retrieved from: https://ceraarchive.dpmc.govt.nz/sites/default/files/Documents/cabinet-minute-choices-for-residents-redzone-crown-offers-july-2011.pdf

CERA (2013). Land Use Recovery Plan *Te Mahere Whakahaumanu Taone*: Summary. Canterbury Earthquake Recovery Authority, Christchurch.

CERA (2014). *Canvas. Your Thinking for the Red Zones. December 2014*. Canterbury Earthquake Recovery Authority, Christchurch.

Chang, S.E., Taylor, J.E., Elwood, K.J., Seville, E., Brunsdon, D. and Gartner, M. (2014). Urban disaster recovery in Christchurch: The central business district cordon and other critical decisions, *Earthquake Spectra* 30 (1), 513–32.

Chatterton, P. (2010). The urban impossible: A eulogy for the unfinished city, *City* 14 (3), 234–44.

Christchurch 360 Trail (2022). The future of the Christchurch 360. Retrieved from: http://christchurch360trail.org.nz/future/

Christchurch City Council (n.d.). Technical Category Information. Retrieved from: https://ccc.govt.nz/consents-and-licences/land-and-zoning/technical-categories-map

Christchurch City Council (2010a). *Christchurch City: Fact Pack 2010*. Strategy and Planning Group, Christchurch City Council, Christchurch. Retrieved from: www.ccc.govt.nz/assets/Documents/Culture-Community/Stats-and-facts-on-Christchurch/fact-packs/FactPack2010-docs.pdf

Christchurch City Council (2010b). *Intensification Community Research Report.* Strategy and Planning Group Community Support Unit, Christchurch, August.

Christchurch City Council (2011). *Central City Plan. Draft Central City Recovery Plan for Ministerial Approval December 2011.* Christchurch City Council, Christchurch.

Christchurch City Council (2013). Christchurch recovery and rebuild issues and challenges: Christchurch city three year plan. Retrieved from: http://resources.ccc.govt. nz/files/ltccp/TYP2013/Volume1/RecoveryRebuildChallenges2013.pdf

Christchurch City Council (2018). *Project 8011.* Retrieved from: https://ccc.govt.nz/ culture-and-community/central-city-christchurch/live-here/residential-programme-8011/

Christchurch City Council (2020). Red zone land returns to local ownership. Retrieved from: https://newsline.ccc.govt.nz/news/story/red-zone-land-returns-to-local-ownership

Christchurch City Council (2021a). Halswell Ward. Christchurch City. Retrieved from: https://ccc.govt.nz/culture-and-community/statistics-and-facts/community-profiles/halswell-hornby-riccarton/halswell-ward

Christchurch City Council (2021b). Ngā Puna Wai sports hub. Retrieved from: https://ccc.govt.nz/rec-and-sport/sports-grounds/nga-puna-wai

Christchurch City Council (2022a). The Enliven Places Programme. Retrieved from: https://ccc.govt.nz/culture-and-community/community-led-development/ enliven-places-programme

Christchurch City Council (2022b). Ōtākaro Avon River Corridor. Co-governance Establishment Committee, Council Supplementary Agenda, 7 April.

Christchurch City Council (2022c). *Draft Annual Plan 2022-23. Activity Plan. Ōtākaro Avon River Corridor (OARC),* Attachments under Separate Cover, 24 February.

Christchurch NZ (2021). *Activities and Attractions.* Retrieved from: www. christchurchnz.com/explore/activities-attractions

Christensen, A.A. (2013). Mastering the land. Mapping and metrologies in Aotearoa New Zealand, in E. Pawson and T. Brooking eds., *Making A New Land. Environmental Histories of New Zealand.* Otago University Press, Dunedin, 310–27.

Chugh, K. (2020). More-than-human cities: On urban design and non-human agency. Retrieved from: https://cuesonline.org/2020/07/26/more-than-human-cities-on-urban-design-and-nonhuman-agency/

Chukwudumogu, I.C. (2018a). The Decision-Making Behaviour of Property Owners (Investors and Developers): A Case Study of the Christchurch City Centre Rebuild. Unpublished PhD thesis, The University of Auckland, Auckland.

Chukwudumogu, I.C. (2018b). The perspectives of commercial property stakeholders in post-disaster rebuild, *Asian Social Science* 14 (5), 82–94.

Chukwudumogu, I.C., Levy, D. and Perkins, H. (2019). The influence of sentiments on property owners in post-disaster rebuild: A case study of Christchurch, New Zealand, *Property Management* 37 (2), 243–61.

Church, L. (2017). Clean-up of Christchurch's 'dirty 30' building list. *Radio New Zealand,* 27 May. Retrieved from: www.rnz.co.nz/news/national/331696/ clean-up-of-christchurch-s-dirty-30-building-list.

CIAL (2021). Financial reports, Christchurch International Airport Company. Retrieved from: www.christchurchairport.co.nz/about-us/who-we-are/financial-reports/

City Scene (2001). Future Path Canterbury, *City Scene Christchurch,* November–December. Retrieved from: http://archived.ccc.govt.nz/CityScene/2001/November/ FuturePathCanterbury.asp

Civil Defence (2020). Declared states of emergency. National Emergency Management Agency. Retrieved from: www.civildefence.govt.nz/resources/previous-emergencies/declared-states-of-emergency/

Clark, H. (2021). Design of new Christchurch indoor stadium confirmed with 25,000 seats, *New Zealand Herald*, 22 July. Retrieved from: www.nzherald.co.nz/nz/design-of-new-christchurch-indoor-stadium-confirmed-with-25000-seats/VVURMDVQDSH5H43U2EHYYTM

Clark, N. (2011). *Inhuman Nature*. Sage, London.

Cloke, P., Beaumont, J. and Williams, A. eds. (2013). *Working Faith: Faith-based Organisations and Urban Social Justice*. Paternoster, Milton Keynes.

Cloke, P. and Conradson, D. (2018). Transitional organisations, affective atmospheres and new forms of being in-common: Post-disaster recovery in Christchurch, New Zealand, *Transactions Institute of British Geographers* 43 (3), 360–76.

Cloke, P. and Dickinson, S. (2019). Transitional ethics and aesthetics. Re-imagining the post-disaster city in Christchurch, New Zealand, *Annals of the Association of American Geographers* 109 (6), 1922–40.

Cloke, P., Dickinson, S. and Tupper, S. (2017). The Christchurch earthquakes 2010, 2011: Geographies of an event, *New Zealand Geographer* 73 (2), 69–80.

Cloke, P. and Johnston, R. eds. (2005). *Spaces of Geographic Thought: Deconstructing Human Geography's Binaries*. Sage, London.

Cloke, P. and Jones, O. (2003). Grounding ethical mindfulness for/in nature: Trees in their places, *Ethics, Place and Environment* 6 (3), 195–213.

Cloke, P. and Pawson, E. (2008). Memorial trees and treescape memories, *Environment and Planning D, Society and Space* 26 (1), 107–22.

Colbert, J., Sila-Nowicka, K. and Yao, J. (2022). Driving forces of population change following the Canterbury Earthquake Sequence, New Zealand: A multiscale geographically weighted regression approach, *Population, Space and Place*, e83. https://doi-org.ezproxy.auckland.ac.nz/10.1002/psp.2583

Conradson, D. (2008). Expressions of charity and action towards justice: Faith-based welfare provision in urban New Zealand, *Urban Studies* 45 (10), 2117–41.

Conradson, D. (2012). Leaving, waiting, staying: Household mobilities after the Christchurch earthquakes. Paper presented to New Zealand Geographical Society Conference, 3–6 December 2012, Napier.

Conway, G. (2011). Christchurch earthquake: Suburbs 'not forgotten', *Stuff*, 3 March. Retrieved from: www.stuff.co.nz/national/christchurch-earthquake/4725553/Christchurch-earthquake-Suburbs-not-forgotten

Conway, G. (2014). Officials ignore local investors, *Stuff*, 16 April. Retrieved from: www.stuff.co.nz/the-press/news/city-centre/9945920/Officials-ignore-local-investors

Coombs, S., Eberlein, A., Mantana, K., Turnhout, A. and Smith, C. (2015). Did dog ownership influence perceptions of adult health and well-being during and following the Canterbury earthquakes? A qualitative study, *Australasian Journal of Disaster and Trauma Studies* 19, 67–75.

Cowlishaw, S. and Mathewson, N. (2012). Officials 'hindered efforts' of volunteers, *Stuff*, 1 March. Retrieved from: www.stuff.co.nz/national/6503366/Officials-hindered-efforts-of-volunteers

Cox, T. (2015). 10 of the world's best concert halls, *The Guardian*, 5 March. Retrieved from: www.theguardian.com/travel/2015/mar/05/10-worlds-best-concert-halls-berlin-boston-tokyo

Cretney, R.M. (2016). Local responses to disaster: The value of community-led post-disaster response action in a resilience framework, *Disaster Prevention and Management* 25 (1), 27–40.

Cretney, R.M. (2017). Towards a critical geography of disaster recovery politics: Perspectives on crisis and hope, *Geography Compass* 11 (1), e12302.

Cretney, R.M. (2018). Beyond public meetings: Diverse forms of community led recovery following disaster, *International Journal of Disaster Risk Reduction* 28, 122–30.

Cretney, R.M. (2019). 'An opportunity to hope and dream'. Disaster politics and the emergence of possibility through community-led recovery, *Antipode* 51 (2), 497–516.

Cretney, R.M. and Bond, S. (2014). Bouncing back to capitalism? Grass-roots autonomous activism in shaping discourses of resilience and transformation following disaster, *Resilience* 2 (1), 18–31.

Cropp, A. (2011). Blue chip earthquake blues, *Stuff*, 13 February. Retrieved from: www.stuff.co.nz/sunday-star-times/features/4647691/Blue-chip-earthquake-blues

Cropper, E. (2017). Desperate Christchurch earthquake victims still fighting for repairs, *Newshub*, 26 June. Retrieved from: www.newshub.co.nz/home/new-zealand/2017/06/desperate-christchurch-earthquake-victims-still-fighting-for-repairs.html

Cubrinovski, M., Green, R.A., Allen, J. et al. (2010). Geotechnical reconnaissance of the 2010 Darfield (Canterbury) earthquake, *Bulletin for the New Zealand Society for Earthquake Engineering* 43 (4), 243–320.

Cupples, J. and Glynn, K. (2009). Countercartographies: (New Zealand) cultural studies/geographies of the city, *New Zealand Geographer* 65 (1), 1–5.

Curtis, A., Mills, J.W. and Leitner, M. (2007). Katrina and vulnerability. The geography of stress, *Journal of Health Care for the Poor and Underserved* 18 (2), 315–30.

Cutter, S., Barnes, L., Berry, M., Burton, C., Evans, E., Tate, R. and Webb, J. (2008). A place-based model for understanding community resilience to natural disasters, *Global Environmental Change* 18 (4), 598–606.

Cutter, S.L., Boruff, B.J. and Shirley, W.L. (2003). Social vulnerability to environmental hazards, *Social Science Quarterly* 84 (2), 242–61.

Cutter, S.L., Emrich, C.T., Mitchell, J.T., Boruff, B.J., Gall, M., Schmidtlein, M.C., Burton, C.G. and Melton, G. (2006). The long road home: Race, class, and recovery from Hurricane Katrina, *Environment: Science and Policy for Sustainable Development* 48 (2), 8–20.

Dann, J. (2014). Those left standing. Retrieved from: https://rebuildingchristchurch.wordpress.com/2014/03/10/those-left-standing/

Dann, J. (2021). The last city of the 20th century, SLATE, 21 February. Retrieved from: https://slate.com/business/2021/02/christchurch-earthquake-anniversary-how-the-rebuild-failed.html

David, L. and Halbert, L. (2014). Finance capital, actor-network theory and the struggle over calculative agencies in the business property markets of Mexico City Metropolitan Region, *Regional Studies* 48 (3), 516–29.

de Bruin, A. and Flint-Hartle, S. (2003). A bounded rationality framework for property investment behaviour, *Journal of Property Valuation and Investment* 21 (3), 271–84.

Department of Building and Housing (2012). *Structural Performance of Christchurch CBD Buildings in the 22 February 2011 Aftershock. Report of an Expert Panel Appointed by the New Zealand Department of Building and Housing*, Wellington.

Department of Statistics (1990). *New Zealand Official 1990 Yearbook*. Department of Statistics, Auckland.

Department of Statistics (2010). *New Zealand Official Yearbook 2010*. David Bateman Ltd, Auckland.

DesRoches, R., Comerio, M. and Eberhard, M. (2011). Overview of the 2010 Haiti earthquake, *Earthquake Spectra* 27 (1, Suppl.), 1–21.

DPMC (2016). *Waimakariri Residential Red Zone Recovery Plan. He Mahere Whakarauora i te Whenua Rāhui o Waimakariri.* Greater Christchurch Group, Department of Prime Minister and Cabinet, Christchurch.

DPMC (2017). *Whole of Government Report. Lessons from the Canterbury Earthquake Sequence.* Department of the Prime Minister and Cabinet, Greater Christchurch Group, Christchurch.

DPMC (2019). Final proposed Global Settlement Agreement with Christchurch City Council, Christchurch, Department of Prime Minister and Cabinet. Retrieved from: https://dpmc.govt.nz/sites/default/files/2019-12/GCR-DEV-19-SUB-0253-4206143. pdf

DeWolfe, D.J. (2000). *Training Manual for Mental Health and Human Service Workers in Major Disasters.* FEMA and Center for Mental Health Services, Washington DC.

Dewsbury, J-D. (2007). Unthinking subjects: Alain Badiou and the event of thought in thinking politics, *Transactions Institute of British Geographers* 32 (4), 443–59.

Diaz, J. III (1999). The first decade of behavioural research in the discipline of property, *Journal of Property Investment and Finance* 17 (4), 326–32.

Dickinson, S. (2013). Post-Disaster Mobilities: Exploring Household Relocation after the Canterbury Earthquakes. Unpublished Master of Science thesis, University of Canterbury, Christchurch.

Dickinson, S. (2018). Spaces of post-disaster experimentation: Agile entrepreneurship and geological agency in emerging disaster countercartographies, *Environment and Planning E: Nature and Space* 1 (4), 621–40.

Dionisio, M.R. (2020). Personal communication. 3 March. Rita Dionisio was then a board member of Te Putahi.

Dionisio, M.R. and Pawson, E. (2016). Building resilience through post-disaster community projects: Responses to the 2010 and 2011 Christchurch earthquakes and 2011 Tokohu tsunami, *Australasian Journal of Disaster and Trauma Studies* 20, 107–16.

Dixon Productions (2011). Christchurch 1996 quake doco – why buildings collapse. Retrieved from: www.youtube.com/watch?v=NkTy6ogLDX8&t=318s

Dodd, F.J. (2021). Exposing the hidden politics of housing provision in Aotearoa New Zealand: The complex governance landscape of new housing in Hamilton. Unpublished PhD thesis, University of Waikato, Hamilton.

Dodds, K. (2020). Staying with trouble, *Territory, Politics and Governance* 8 (2), 139–43.

Dombroski, K. (2020). Personal communication, 5 March. Kelly Dombroski was then a board member of Life in Vacant Spaces.

Dombroski, K., Diprose, G. and Boles, I. (2019). Can the commons be temporary? The role of transitional commoning in post-quake Christchurch, *Local Environment* 24 (4), 313–28.

Dominey-Howes, D., Gorman-Murray, A. and McKinnon, S. (2013). Queering disasters: On the need to account for LGBTI experiences in natural disaster contexts, *Gender, Place and Culture* 21 (7), 905–18.

Dorahy, M.J. and Kannis-Dymand, L. (2012). Psychological distress following the 2010 Christchurch earthquake: A community assessment of two differentially affected suburbs, *Journal of Loss & Trauma* 17 (3), 203–17.

Doudney, T. (2015). Expert warns of urban sprawl. *The Star*, 22 April, 15.

Dwyer, S. and Buckle, J. (2009). The space between: On being an insider-outsider in qualitative research, *International Journal of Qualitative Methods* 8 (1), 54–63.

Easton, B. (2000). Such, William Ball, 1907–75, *Dictionary of New Zealand Biography, Volume Five, 1941–60*. Auckland University Press, Auckland and Department of Internal Affairs, Wellington, 504–6.

Edmunds, S. (2017). 'Over-enthusiastic investment by unsophisticated people', *Stuff*, 15 September. Retrieved from: www.stuff.co.nz/business/96888627/overenthusiastic-investment-by-unsophisticated-people?rm=a

Elder, D.M., McCahon, I.F. and Yetton, M.D. (1991). *The Earthquake Hazard in Christchurch. A Detailed Evaluation*. Earthquake Commission, Wellington.

Enarson, E. and Morrow, B.H. eds. (1988). *The Gendered Terrain of Disaster*. Greenwood, Westport CT.

Enarson, E. and Pease, B. eds. (2016). *Men, Masculinities and Disaster*. Routledge, Abingdon.

England, K. (1994). Getting personal: Reflexivity, positionality and feminist research, *The Professional Geographer* 46 (1), 80–9.

Ensor, B. (2011). Remarkable effort by Student Army, *Stuff*, 25 February. Retrieved from: www.stuff.co.nz/marlborough-express/4705214/Remarkable-effort-by-student-army

Environment Canterbury (2011). *Ecological Effects of the Christchurch February Earthquake on our City Estuary*. Retrieved from: www.eosecology.co.nz/files/Ecan-Quake-Estuary-Ecology.pdf

Environment Canterbury (2013). *Natural Environment Recovery Programme for Greater Christchurch, Whakaara Taiao*, Report Number R13/68, Canterbury Regional Council, Christchurch.

Environment Canterbury (2021a). *How Many People Live in Canterbury?* Retrieved from: www.ecan.govt.nz/your-region/living-here/regional-leadership/population/census-estimates/

Environment Canterbury (2021b). Greater Christchurch Metro. Retrieved from: www.ecan.govt.nz/your-region/living-here/transport/public-transport-services/greater-christchurch-metro/

EPIC (2022). EPIC Innovation. Retrieved from: https://epicinnovation.co.nz

EQC (2021). Drainage claims. Earthquake Commission. Retrieved from: https://www.eqc.govt.nz/insurance-and-claims/canterbury-earthquake/canterbury-claims/drainage-claims/

Ermacora, T. and Bullivant, L. (2016). *Recoded City: Co-creating Urban Futures*, Routledge, London.

Erskine, S. (2022). Personal communication, 31 January. Shelley Erskine was then marketing manager at the Christchurch Arts Centre.

Farrell, F. (2012). *The Broken Book*. Auckland University Press, Auckland.

Farrell, F. (2015). *Villa at the Edge of the Empire. One Hundred Ways to Read a City*. Penguin Random House, Auckland.

Feeney, W. (2019). *Out There: SCAPE Public Art, 1998–2018*. SCAPE, Christchurch.

Finlayson, C. (2021). Ōtākaro/Avon River Corridor co-governance options. Letter to the Mayor of Christchurch, 21 June.

Finlayson, C. and Christmas, J. (2021). *He Kupu Taurangi. Treaty Settlements and the Future of Aotearoa New Zealand*. Huia Publishers, Wellington.

Fisher, M. (2020). Personal communication, 27 February. Mike Fisher was then general manager of Christchurch's Riverside Market, director of UrbanTacticians and chair of the board of Te Putahi.

Forer, P. (1978). Time-space and area in the city of the plains, in T. Carlstein, D. Parkes and N. Thrift eds., *Timing Space and Spacing Time*. Edward Arnold, London, 99–118.

Franklin, A. (2017). The more-than-human city, *The Sociological Review* 65 (2), 202–17.

Fullilove, M. (2016). *Root Shock. How Tearing Up City Neighborhoods Hurts America, and What We Can Do About It*. NYU Press, New York.

Fusté-Forné, F. (2017). Building experiencescapes in Christchurch, *Landscape Review (Lincoln)* 17 (1), 44–57.

Gablik, S. (1992). The ecological narrative, *Art Journal* 51 (2), 49–51.

Gablik, S. (1996). Connective aesthetics: Art after individualism, in S. Lacy ed., *Mapping the Terrain: New Genre Public Art*. Bay Press, Seattle WA, 74–87.

Gaillard, J.C., Gorman-Murray, A. and Fordham, M. (2017). Sexual and gender minorities in disaster. *Gender, Place and Culture* 24 (1), 18–26.

Gallagher, K. and Pyle, D. (2021). *Rohe Kōreporepo. The Swamp, the Sacred Place*. Wick Candle, Christchurch.

Galloway, M. (2014). A message and a messenger, in B. Bennett, J. Dann, E. Johnson and R. Reynolds eds., *Once in a Lifetime: City-building after Disaster in Christchurch*, Freerange Press, Christchurch, 112–16.

Galloway, M. (2021). Art over nature over art. (Re)imagining Ōtautahi Christchurch, in F. Freschi, J. Venis and F. Nazier eds., *The Politics of Design. Privilege and Prejudice in Aotearoa New Zealand, Australia and South Africa*. Otago Polytechnic Press, Dunedin, 275–91.

Gap Filler (2022). Placemaking at One Central. Retrieved from: https://gapfiller.org.nz/project/placemaking-at-one-central/

Garces-Ozanne, A., Makabenta-Ikeda, M. and Uekusa, S. (2022). Asian migrant worker experiences in Ōtautahi Christchurch, in S. Uekusa, S. Matthewman and B.C. Glavovic eds., *A Decade of Disaster Experiences in Ōtautahi Christchurch*. Palgrave Macmillan, Singapore, 211–36.

Gardner, R.O. (2013). The emergent organization: Improvisation and order in Gulf Coast disaster relief, *Symbolic Interaction* 36 (3), 237–60.

Gates, C. (2012a). Central city abuzz with noise of demolition, *The Press*, 29 September, 1.

Gates, C. (2012b). Dalziel claims CERA ditched her, *Stuff*, 24 April. Retrieved from: www.stuff.co.nz/the-press/6795039/Dalziel-claims-Cera-ditched-her

Gates, C. (2015a). 1240 central Christchurch buildings demolished, *Stuff*, 20 February. Retrieved from: www.stuff.co.nz/the-press/news/christchurch-earthquake-2011/662 90638/1240-central-christchurch-buildings-demolished

Gates, C. (2015b). St Elmos Court owner bitter, angry, in debt, *Stuff*, 20 February. Retrieved from: www.stuff.co.nz/the-press/news/christchurch-earthquake-2011/66248945/st-elmos-court-owner-bitter-angry-in-debt

Gates, C. (2015c). Fruit foraging in Christchurch's red zone, *Stuff*, 10 April. Retrieved from: www.stuff.co.nz/life-style/food-wine/food-news/67690529/fruit-foraging-in-christchurchs-red-zone

Gates, C. (2018). Empty central Christchurch sites cover land three times the size of Botanic Gardens, *Stuff*, 8 January. Retrieved from: www.stuff.co.nz/the-press/business/the-rebuild/100390195/empty-central-christchurch-sites-cover-land-three-times-the-size-of-botanic-gardens

Gates, C. (2021). Breeding colony for rare gulls in central Christchurch ruins is being demolished, *Stuff*, 18 July. Retrieved from: www.stuff.co.nz/the-press/news/125738537/breeding-colony-for-rare-gulls-in-central-christchurch-ruins-is-being-demolished

Gehl, J. (2015). In search of the human scale, TEDx talk, 19 December. Retrieved from: www.youtube.com/watch?v=Cgw9oHDfJ4k

Gehl Architects (2010). *Christchurch 2009: Public Space, Public Life*. Christchurch City Council.

Ghosh, A. (2021). *The Nutmeg's Curse. Parables for a Planet in Crisis*. John Murray, London.

Gibson, A. (2019). Four biggest NZ malls a focus for Scentre in $790m property expansion, *New Zealand Herald*, 11 April. Retrieved from: www.nzherald.co.nz/business/four-biggest-nz-malls-a-focus-for-scentre-in-790m-property-expansion/YMCZXREWZMLMWO74SR7GZ3UEJU/

Giovinazzi, S., Wilson, T., Davis, C., Bristow, D., Gallagher, M., Schofield, A., Villemure, M., Eidinger, J. and Tang, A. (2014). Lifelines performance and management following the 22 February 2011 Christchurch earthquake, New Zealand: Highlights of resilience, *Bulletin of the New Zealand Society for Earthquake Engineering* 44 (4), 402–17.

Gjerde, M. (2017a). A change of scene: How the Canterbury earthquake sequence led to a departure from concrete technologies, in P. Chan and C. Neilson eds., *Proceedings of the 33rd Annual Association of Researchers in Construction Management Conference, Cambridge*, 4–6 September 2017, 724–33.

Gjerde, M. (2017b). Building back better: Learning from the Christchurch rebuild, *Procedia Engineering* 198, 530–40.

Glaeser, E. and Sassen, S. (2011). 16 global cities to watch, *Foreign Policy*, December, 72 and 85.

Glavovic, B. (2022). IPCC Report: Coastal cities are sentinels for climate change. It's where our focus should be as we prepare for inevitable impacts, *The Conversation*, 1 March.

Global Settlement Agreement (2019). *Global Settlement Agreement, 23 September 2019*, Christchurch City Council and the Crown. Retrieved from: www.ccc.govt.nz/assets/Documents/The-Council/Plans-Strategies-Policies-Bylaws/Strategies/Global-Settlement/CCC-Release-Global-Settlement-Agreement-23-Septmeber-2019.pdf

Gluckman, P. (2011). The psycho-social consequences of the Canterbury earthquakes. A briefing paper. Office of the Prime Minister's Science Advisory Committee, 10 May.

Gordon, E., Sutherland, J., Du Plessis, R. and Gibson, H. (2014). *Movers and Shakers: Women's Stories from the Christchurch Earthquakes*. Women's Voices Project - Ngā Reo O Ngā Wahine. Retrieved from: www.communityresearch.org.nz/wp-content/uploads/formidable/MOVERS-AND-SHAKERS_Final.pdf

Gorman, P. (2010a). Scientists can't rule out fault below Chch, *The Press*, 6 October.

Gorman, P. (2010b). The science behind the shakes, *Stuff*, 26 December. Retrieved from: www.stuff.co.nz/the-press/news/christchurch-earthquake-2011/canterbury-earthquake-2010/4495446/The-science-behind-the-shakes

Gorman, P. (2016). *Portacom City: Reporting on the Christchurch and Kaikōura Earthquakes*. Bridget William Books, Wellington.

Gray, R. (2016). Monuments or millstones, in J. Harvey ed., *Christchurch – Five Years On, Architecture New Zealand*, March/April, 45–7.

Greater Christchurch Urban Development Strategy (2007). *Greater Christchurch Urban Development Strategy 2007*. Retrieved from: https://greaterchristchurch.org.nz/assets/Documents/greaterchristchurch/UDSActionPlan2007.pdf

Greater Christchurch Urban Development Strategy (2016). *Greater Christchurch Urban Development Strategy Update 2016*, Canterbury Regional Council, Christchurch.

Greater Christchurch Partnership (2018). *Our Space 2018-2048: Greater Christchurch Settlement Pattern Update Whakahāngai O Te Hōrapa Nohoanga*. Draft for Consultation, Christchurch, November.

Greater Christchurch Urban Development Strategy Implementation Committee (2011). Committee minutes. Retrieved from: http://archived.ccc.govt.nz/council/agendas/2011/february/greaterchchudsic21st/gcudsic21february2011-fullagenda.pdf

Greaves, L.M., Milojev, P., Huang, Y., Stronge, S., Osborne, D., Bulbulia, J., Grimshaw, M. and Sibley, C.G. (2015). Regional differences in the psychological recovery of Christchurch residents following the 2010/2011 earthquakes: A longitudinal study, *PLOS One*, 10 (5), e0124278.

Green Lab (2022). Linwood tiny shops village. Retrieved from: https://thegreenlab.co.nz/linwood-tiny-shops-village/; https://www.tellinglives.co.nz/tiny-shops-village

Greene, B. (2021). *You Can't Rush an Earthquake. Repair, Rebuild, Repeat.* Brenda Greene, Christchurch.

Greenhill, M. (2013). Rebuilds under way on tricky TC3 sites, *Stuff*, 11 July. Retrieved from: www.stuff.co.nz/the-press/8903528/Rebuilds-under-way-on-tricky-TC3-sites

Greenhough, B. (2010). Vitalist geographies: Life and the more-than-human, in P. Harrison and B. Anderson eds., *Taking Place: Non-representational Theories and Geography*. Ashgate, London, 37–54.

Grimshaw & Co (2014). Poor repairs to haunt buyers. Retrieved from: www.grimshaw.co.nz/news-resources/poor-repairs-to-haunt-buyers/

Grimshaw, M. (2015). Personal communication, 4 March. Mike Grimshaw was then Associate Professor of Sociology and Anthropology at the University of Canterbury.

Grollimund, B. (2014). *Small Quake, Big Impact: Lessons Learned from Christchurch*. Swiss Re, Zurich.

Gruen, L. (2015). *Entangled Empathy. An Alternative Ethic for our Relationship with Animals*. Lantern Books, Brooklyn.

Gundermann, B. (2014). *A Holistic Transition to Climate-resilient Cities*. Urbia Group, Auckland.

Hackworth, J. (2007). *The Neoliberal City. Governance, Ideology, and Development in American Urbanism*. Cornell University Press, Ithaca.

Halbert, L., Henneberry, J. and Mouzakis, F. (2014). The financialization of business property and what It means for cities and regions, *Regional Studies* 48 (3), 547–50.

Hall, C.M., Malinen, S., Vosslamber, R. and Wordsworth, R. eds. (2016). *Business and Post-disaster Management Business, Organisational and Consumer Resilience and the Christchurch Earthquakes*. Routledge, London.

Hall, P. (1982). *Great Planning Disasters*. University of California Press, Berkeley.

Hamilton, C. (2016). Fun for all ages. *Australasian Leisure Management*. September/October, 18–22.

Haraway, D. (1988). Situated knowledges: The science question in feminism and the privilege of partial perspective, *Feminist Studies* 14 (3), 575–99.

Haraway, D.J. (2016). *Staying with the Trouble. Making Kin in the Chthulucene*. Duke University Press, Durham.

Hare, J. (2011). Preliminary observations from Christchurch earthquakes. Technical report, Structural Engineering Society of New Zealand. Retrieved from: https://canterbury.royalcommission.govt.nz/documents-by-key/2011-09-23108

Harris, D. and Hayward, M. (2019). MPs urge tighter house-building rules to protect neighbourhoods. *Stuff*, 18 March. Retrieved from: www.stuff.co.nz/the-press/ news/110931729/mps-urge-planning-law-changes-to-protect-against- housing-intensification?rm=a

Harvey, D. (1989a). *The Condition of Postmodernity. An Enquiry into the Origins of Cultural Change.* Basil Blackwell, Oxford.

Harvey, D. (1989b). From managerialism to entrepreneurialism: The transformation of urban governance in late capitalism, *Geografiska Annaler* 71 B (1), 3–17.

Harvey, S. (2012). The Grace Vineyard Church response, *Tephra* 23, 8–12.

Harvey, S. (2014). Sam's story, in M. Parsons ed., *Rubble to Resurrection: Churches Respond in the Canterbury Quakes.* DayStar, Auckland, 223–4.

Harvie, W. (2016). How Māori will rebuilt Christchurch look? *Stuff*, 23 September. Retrieved from:www.stuff.co.nz/the-press/business/the-rebuild/84195996/how-maori- will-rebuilt-christchurch-look

Harvie, W. (2021). Environmental groups at odds over Christchurch estuary walkway, *Stuff*, 8 August. Retrieved from: www.stuff.co.nz/the-press/news/125964428/ environmental-groups-at-odds-over-christchurch-estuary-walkway

Hawkes, C. (2018). Elite streets: Christchurch wants the leafy, established avenues in Fendalton, *Stuff*, 19 March. Retrieved from: www.stuff.co.nz/life-style/homed/ latest/101391044/christchurchs-elite-streets-are-the-leafy-established-avenues-in- fendalton

Hawkey, J. (2018). Personal communication, 5 November. Jill Hawkey was Executive Director of Christchurch Methodist Mission.

Hawkey, J. (2020). Personal communication, 18 March. Jill Hawkey was Executive Director of Christchurch Methodist Mission.

Hayward, B. (2012). Canterbury's political quake, *Stuff*, 30 March. Retrieved from: www.stuff.co.nz/the-press/opinion/perspective/6664104/Canterburys-political-quake/

Hayward, B.M. (2013). Rethinking resilience: Reflections on the earthquakes in Christchurch, New Zealand, 2010 and 2011, *Ecology and Society* 18 (4), 37.

Hayward, B. and Cretney, R. (2015). Governing through disaster, in J. Hayward ed., *New Zealand Government and Politics*, 6th edn., Oxford University Press, Melbourne, 403–15.

Hayward, M. (2018). Nearly seven years on, thousands of Christchurch earthquake insurance claims remain, *Stuff*, 10 February. Retrieved from: www.stuff.co.nz/life- style/homed/latest/101391044/christchurchs-elite-streets-are-the-leafy- established-avenues-in-fendalton

Hayward, M. (2020a). Plan for unique central city neighbourhoods to bring people into the city, *Stuff*, 1 August. Retrieved from: www.stuff.co.nz/the-press/ news/122306971/plan-for-unique-central-christchurch-neighbourhoods-to- bring-people-into-the-city

Hayward, M. (2020b). Christchurch's Riverside Market thriving after a year in business, *Stuff*, 2 October. Retrieved from: www.stuff.co.nz/business/our-backyard/ 122965722/christchurchs-riverside-market-thriving-after-a-year-in-business

Hayward, M. and McDonald, L. (2019). Fancy plans for Cathedral Square won't take shape for at least a decade, *Stuff*, 19 January. Retrieved from: www.stuff.co.nz/ business/110025110/fancy-plans-for-cathedral-square-wont-take-shape-for-at- least-a-decade

Haworth, G. (2019). *Guts and Grace. Christchurch City Mission, A History*. Wily, Christchurch.

He, L., Dominey-Howes, D., Aitchinson, J.C., Lau, A. and Conradson, D. (2021). How do post-disaster policies influence household-level recovery? A case study of the 2010–11 Canterbury earthquake sequence, New Zealand, *International Journal of Disaster Risk Reduction* 60, 102274.

Heather, B. (2011a). Chemical toilet confusion in earthquake-hit Christchurch, *Stuff*, 31 March. Retrieved from: www.stuff.co.nz/national/christchurch-earthquake/483 1055/Chemical-toilet-confusion-in-earthquake-hit-Christchurch

Heather, B. (2011b). Red zone house owners reject buyout, *Stuff*, 26 September. Retrieved from: www.stuff.co.nz/national/5682708/Red-zone-house-owners-reject-buyout

Heins, A. (2015). Greenfield Drivers: Residential Choice, Transport, and the Subjective Experience of Travel in the Greenfield Suburbs – A Case Study in Christchurch. Unpublished Master of Science thesis, University of Canterbury, Christchurch.

Herman, J.L. (1998). Recovery from psychological trauma, *Psychiatry and Clinical Neurosciences* 52 (S1), S98–103.

Hewitt, K. (1997). *Regions of Risk. A Geographical Introduction to Disasters.* Longman, Harlow.

Hinchliffe, S. and Whatmore, S. (2006). Living cities: Towards a politics of conviviality, *Science as Culture* 15 (2), 123–38.

Hodder, I. (2014). The entanglement of humans and things: A long term view, *New Literary History* 45 (1), 19–36.

Hogg, D., Kingham, S., Wilson, T.M., Griffin, E. and Ardagh, M. (2014). Geographic variation of clinically diagnosed mood and anxiety disorders in Christchurch after the 2010/11 earthquakes, *Health and Place* 30, 270–78.

Hörhager, E. (2015). Political implications of natural disasters: Regime consolidation and political contestation, *WIT Transactions on the Built Environment* 150, 271–81.

Hou, L. and Shi, P. (2011). Haiti 2010 earthquake - how to explain such huge losses? *International Journal of Disaster Risk Science* 2 (1), 25–33.

How, S.M. and Kerr, G.N. (2019). Earthquake impacts on immigrant participation in the Greater Christchurch construction labor market, *Population Research and Policy Review* 38 (2), 241–69.

Huck, A., Monstadt, J. and Driessen, P. (2020). Building urban and infrastructure resilience through connectivity: An institutional perspective on disaster risk management in Christchurch, New Zealand, *Cities* 98, 102573.

Hughes, R. (1991). *The Shock of the New. Art and the Century of Change.* Thames & Hudson, London.

Hughes, M.W., Quigley, M.C., van Ballegooy, S., Deam, B.L., Bradley, B.A., Hart, D.E. and Measures, R. (2015). The sinking city: Earthquakes increase flood hazard in Christchurch, New Zealand, *GSA Today* 25 (3–4), 4–10.

Hunt, E. (2021a). 'Joy and agony': Christchurch earthquake survivors ten years on. *The Guardian*, 19 February. Retrieved from: www.theguardian.com/world/2021/feb/20/trauma-and-transformation-christchurch-earthquake-survivors-ten-years-on

Hunt, E. (2021b). Wings but no prayers: Seagulls delay rebuilding of Christchurch cathedral, *The Guardian*, 10 February. Retrieved from: www.theguardian.com/world/2021/feb/11/worlds-rarest-seagulls-thrive-in-demolished-new-zealand-office-block

Ineson, J. (2018). Rebuilt QEII centre puts east Christchurch 'back on the map' as hundreds stream in on opening night, *Stuff*, 31 May. Retrieved from: www.stuff.co.nz/the-press/news/104383364/rebuilt-qeii-centre-puts-east-christchurch-back-on-the-map-as-hundreds-stream-in-on-opening-night

Insurance Council of New Zealand (2021). *Canterbury Earthquakes.* Retrieved from: www.icnz.org.nz/natural-disasters/canterbury-earthquakes/

International Labour Organisation (2021). *Haiti Earthquake 2021: Post Disaster Needs Assessment*. Retrieved from: www.ilo.org/global/docs/wcms_831127/lang--en/index.htm

IPCC (2022). *Climate Change 2022. Impacts, Adaptation and Vulnerability*, Intergovernmental Panel on Climate Change. Retrieved from: www.ipcc.ch

IPWEA. (2019). Australasian Award for Margaret Mahy Family Playground. Retrieved from: www.wsp.com/en-NZ/news/2019/australasian-award-for-margaret-mahy-family-playground

Irvine, R.D.G. (2017). Anthropocene East Anglia, *The Sociological Review Monographs* 65 (1), 154–70.

Jackson, M. (2021). Where to next? Decolonisation and the stories in the land, in B. Elkington, M. Jackson, R. Kiddle, Mercier, O.R., Ross, M., Smeaton, J. and Thomas, A. *Imagining Decolonisation*. Bridget Williams Books, Wellington, 133–55.

Jacobs, J. (1961). *The Death and Life of Great American Cities*. Random House, New York.

Jamieson, A. (2018). Personal communication, 2 November. Alan Jamieson was Senior Pastor of South West Baptist Church.

Jeffrey, C. and Dyson, J. (2021). Geographies of the future: Prefigurative politics, *Progress in Human Geography* 45 (4), 641–58.

Jewell, N. (2014). Earthquake resistant eco-village wins Christchurch's Breathe competition, *Inhabitat*. Retrieved from: https://inhabitat.com/dynamic-team-comprised-of-local-builders-and-italian-architects-wins-christchurchs-breathe-competition/

Johnson, L.A. and Olshansky, R.B. (2017). *After Great Disasters. An In-Depth Analysis of How Six Countries Managed Community Recovery*. Lincoln Institute of Land Policy, Cambridge MA.

Johnson, M.L. (2006). Geographical reflections on the 'new' New Orleans in the post-Hurricane Katrina era, *Geographical Review* 96 (1), 139–56.

Johnson, S. (2012). Students vs. the machine, *Tephra* 23, 18–22.

Johnston, K.M. (2014). Planning for a Night Out: Local Governance, Power and Night-Time in Christchurch, New Zealand. Unpublished PhD thesis, Massey University, Palmerston North.

Jones, N. (2015). Immigration rules changed to clear way for 5000 more Christchurch rebuild workers, *New Zealand Herald*, 13 May. Retrieved from: www.nzherald. co.nz/nz/immigration-rules-changed-to-clear-way-for-5000-more-christchurch-rebuild-workers/CBCDJPO4Y7D2IQYKJKHZMPQUQQ/

Jones, O. and Cloke, P. (2002). *Tree Cultures. The Place of Trees and Trees in their Place*. Berg, Oxford.

Joseph, J. (2013a). Resilience as embedded neoliberalism: A governmentality approach, *Resilience* 1 (1), 38–52.

Joseph, M. (2013b). *Fluid New York. Cosmopolitan Urbanism and the Green Imagination*. Duke University Press, Durham.

Judt, T. (2010). *Postwar. A History of Europe since 1945*. Vintage Books, London.

Kahn, M.E. (2005). The death toll from natural disasters: The role of income, geography, and institutions, *The Review of Economics and Statistics* 87 (2), 271–84.

Kaiser, A., Holden, C., Beavan, J., Beetham, D., Benites, R., Celentano, A., Collett, D., Cousins, J., Cubrinovski, M., Dellow, G. and Denys, P. (2012). The M_w 6.2 Christchurch earthquake of February 2011: Preliminary report, *New Zealand Journal of Geology and Geophysics* 55 (1), 67–90.

Kam, W.Y., Pampanin, S. and Elwood, K. (2011). Structural performance of reinforced concrete buildings in the 22 February Christchurch (Lyttelton) earthquake, *Bulletin of the New Zealand Society for Earthquake Engineering* 44 (4), 239–78.

Kane, R. and Smith, J. (2013). Inner City East – one Christchurch community's story, *Aotearoa New Zealand Social Work* 25 (2), 90–7.

Kates, R.W., Colten, C.E., Laska, S. and Leatherman, S.P. (2006). Reconstruction of New Orleans after Hurricane Katrina. A research perspective, *Proceedings of the National Academy of Sciences* 103 (40), 14653–60.

Katz, C. (2008). Bad elements: Katrina and the scoured landscape of social reproduction, *Gender, Place and Culture: A Journal of Feminist Geography* 15 (1), 15–29.

Kenny, J. (2021a). Popular Christchurch bar Viaduct finally returns after post-quake hiatus. *Stuff*, 27 January. Retrieved from: www.stuff.co.nz/the-press/christchurch-life/eat-and-drink/124057522/popular-christchurch-bar-viaduct-finally-returns-after-postquake-hiatus

Kenny, L. (2020). Christchurch's new hot saltwater pools sell-out for opening weekend, *Stuff*, 31 May. Retrieved from: www.stuff.co.nz/travel/back-your-backyard/121679008/christchurchs-new-hot-saltwater-pools-sellout-for-opening-weekend

Kenny, L. (2021b). New pool for east Christchurch after decade-long wait 'makes our hearts happy', *Stuff*, 1 October. Retrieved from: www.stuff.co.nz/the-press/news/126554658/new-pool-for-east-christchurch-after-decadelong-wait-makes-our-hearts-happy

Kenney, C.M. and Phibbs, S. (2015). A Māori love story: Community-led disaster management in response to the Ōtautahi (Christchurch) earthquakes as a framework for action, *International Journal of Disaster Risk Reduction* 14 (1), 46–55.

Kenney, C.M., Phibbs, S.R., Paton, D., Reid, J. and Johnston, D.M. (2015). Community-led disaster risk management. A Māori response to Ōtautahi (Christchurch) earthquakes, *Australasian Journal of Disaster and Trauma Studies* 19, 9–20.

Key, J. (2015). Speech to the Canterbury Employers' Chamber of Commerce lunch, 2 July. Retrieved from: www.stuff.co.nz/the-press/news/69903502/john-keys-speech-to-the-canterbury-employers-chamber-of-commerce-lunch

King, A., Middleton, D., Brown, C., Johnston, D. and Johal, S. (2014). Insurance: Its role in recovery from the 2010–2011 Canterbury earthquake sequence, *Earthquake Spectra* 30 (1), 475–91.

Klein, N. (2008). *The Shock Doctrine*. Penguin, Harmondsworth.

Lambert, S. (2014). Māori and the Christchurch earthquakes. The interplay between Indigenous endurance and resilience through urban disaster, *MAI Journal* 3 (2), 165–80.

Land Information New Zealand (2021). *Pest Animal Survey Results*, Land Information New Zealand. Retrieved from: www.linz.govt.nz/crown-property/types-crown-property/christchurch-residential-red-zones/pest-animal-survey-results

Landscape Architecture Aotearoa (2022). Embedding stories of mana whenua in post-quake Christchurch. Retrieved from: www.landscapearchitecture.nz/landscape-architecture-aotearoa/2022/2/12/debbie-tikao-working-to-embed-the-stories-of-mana-whenua-in-post-quake-christchurch

Latour, B. (2018). *Down to Earth. Politics in the New Climatic Regime*. Polity Press, Cambridge.

Law, T. (2012). Shock at schools shake-up, *Stuff*, 18 September. Retrieved from: www.stuff.co.nz/the-press/news/7674258/Shock-at-schools-shake-up

Law, T. (2015). Work could start soon on restoring the Christchurch Town Hall, *Stuff*, 7 June. Retrieved from: www.stuff.co.nz/the-press/business/the-rebuild/69167736/work-could-start-soon-on-restoring-the-christchurch-town-hall

Law, T. (2016). New pool in Christchurch's east urgently needed, principals say, *Stuff*, 18 May. Retrieved from: www.stuff.co.nz/the-press/news/80117823/new-pool-in-christchurchs-east-urgently-needed-principals-say

Law, T. (2018). Excitement surrounds the opening of Christchurch's long-awaited QEII complex, *Stuff*, 28 May. Retrieved from: www.stuff.co.nz/the-press/news/104259359/excitement-surrounds-the-opening-of-christchurchs-longawaited-qeii-complex

Law, T. (2019a). Christchurch's rubble mountain: The end of an earthquake-demolition dump, *Stuff*, 29 December. Retrieved from: www.stuff.co.nz/the-press/news/118153213/christchurchs-rubble-mountain-the-end-of-an-earthquakedemolition-dump

Law, T. (2019b). Riverside Market brings the 'buzz' back into Christchurch's central city, *Stuff*, 30 September. Retrieved from: www.stuff.co.nz/the-press/news/116198681/riverside-market-brings-the-buzz-back-into-christchurchs-central-city

Law, T. (2020a). New rule could stop new central Christchurch apartments being used for Airbnb, *Stuff*, 11 June. Retrieved from: www.stuff.co.nz/the-press/news/121782640/new-rule-could-stop-new-central-christchurch-apartments-being-used-for-airbnb

Law, T. (2020b). New Christchurch apartments criticised for 'poor' urban design, *Stuff*, 9 September. Retrieved from: www.stuff.co.nz/the-press/news/122695983/new-christchurch-apartments-criticised-for--poor-urban-design

Law, T. (2020c). Council cracks down on 120 unconsented central Christchurch car parks, *Stuff*, 2 December. Retrieved from: www.stuff.co.nz/the-press/news/123571925/council-cracks-down-on-120-unconsented-central-christchurch-car-parks

Law, T. (2021). Christchurch ratepayers urged to support $5.5 m Arts Centre grant, *Stuff*, 21 February. Retrieved from: www.stuff.co.nz/the-press/news/124285896/christchurch-ratepayers-urged-to-support-55m-arts-centre-grant

Law, T. (2022). At-risk gulls find new home, *Stuff*, 25 January. Retrieved from: www.stuff.co.nz/the-press/news/127577553/atrisk-gulls-find-new-home--mistaking-halfbuilt-stormwater-basin-for-a-braided-river

Law, T. and McDonald, L. (2020). Time to turn the heat up: Call for action as 20 per cent of central Christchurch sites still vacant, *The Press*, 10 December, 1.

Le Heron, R. (2018). Dairying in question, in E. Pawson and the Biological Economies Research Team, *The New Biological Economy. How New Zealanders are Creating Value from the Land*. Auckland University Press, Auckland, 20–40.

Le Heron, R. and Pawson, E. eds. (1996). *Changing Places. New Zealand in the Nineties*. Longman Paul, Auckland.

Lesniak, M. (2016). Rebuilding/Reimagining Post-disaster Christchurch. Unpublished Masters Altevilles dissertation, University of Lyon, Lyon.

Levy, D., Hills, R., Perkins, H.C., Mackay, M., Campbell, M. and Johnston, K. (2021). Local benevolent property development entrepreneurs in small town regeneration. *Land Use Policy* 108, 105546.

Lewis, J.N. (2011). The future of network governance research: Strength in diversity and synthesis, *Public Administration* 89 (4), 1221–34.

Liberty, K., Tarren-Sweeney, M., Macfarlane, S., Basu, A. and Reid, J. (2016). Behavior problems and post-traumatic stress symptoms in children beginning school: A comparison of pre- and post-earthquake groups, *PLoS Currents* 8.

Life in Vacant Spaces (n.d.). The East x East project. Retrieved from: https://livs.org.nz/our-spaces/east-x-east/

Lin II, R-G. and Allen, S. (2011). New Zealand quake raises questions about LA buildings, *Los Angeles Times*, 26 February. Retrieved from: www.latimes.com/local/la-me-quake-california-20110226-story.html

Lindblom, C. (1959). The science of muddling through, *Public Administration Review* 19, 79–88.

Lindblom, C.E. (1979). Still muddling, not yet through, *Public Administration Review* 39 (6), 517–26.

LINZ (2015). *Geotechnical Information on Horizontal Land Movement due to the Canterbury Earthquake Sequence*, prepared for Land Information New Zealand by Tonkin and Taylor, Christchurch.

Lochhead, J. ed. (2019). *The Christchurch Town Hall 1965–2019. A Dream Renewed*, Canterbury University Press, Christchurch.

Lonely Planet (2017). *Oi You!* Retrieved from: www.streetart.co.nz/blog/2017/9/23/oi-you-the-driving-force-behind-getting-christchurch-and-adelaide-being-recognised-as-street-art-cities

Longley, G. (2015). Hagley Oval redevelopment ends up costing $4.5m - three times the original estimate, *Stuff*, 22 May. Retrieved from: www.stuff.co.nz/sport/cricket/68779977/hagley-oval-redevelopment-ends-up-costing-45m---three-times-the-original-estimate

Loughrey, D. (2019). A decade on, stadium scars remain, *Otago Daily Times*, 16 February. Retrieved from: www.odt.co.nz/news/dunedin/decade-stadium-scars-remain

Lovatt, A. and O'Connor, J. (1995). Cities and the night-time economy, *Planning Practice and Research* 10 (2), 127–34.

Lowenthal, D. (2004). The heritage crusade and its contradictions, in M. Page and R. Mason eds., *Giving Preservation a History. Histories of Historic Preservation in the United States*. Routledge, New York, 19–43.

Lucas, D. (2014). Does the Blueprint support the creation of a healthy ecological urban environment? in B. Bennett, J. Dann, E. Johnson and R. Reynolds, eds., *Once in a Lifetime: City-building after Disaster in Christchurch*. Freerange Press, Christchurch, 438–49.

Lunday, J. (2020). In G. Smyth, dir., *When A City Rises*, Frank Films. Retrieved from: www.nzonscreen.com/title/when-a-city-rises-2020

Lynch, C. (2019). Has the east of Christchurch been forgotten about? Retrieved from: www.youtube.com/watch?v=83ZnkOVOiDI

Lynch, C. (2020). Election results 2020: Gerry Brownlee defeat one of biggest upsets for National party. Retrieved from: www.nzherald.co.nz/nz/election-results-2020-chris-lynch-gerry-brownlee-defeat-one-of-biggest-upsets-for-national-party/RAH3S47JAW5LG53I4EORHKF4HQ/

Lynch, D. (2014). Response to criticism of Crowne Plaza deputation. Retrieved from: www.facebook.com/GapFiller/posts/866575836703253

MacDonald, M. and Carlton, S. (2016). *Staying in the Red Zones. Monitoring Human Rights in the Canterbury Earthquake Recovery*. Human Rights Commission, Auckland.

MacPherson, R. (2016). Placemaking in Transitional Christchurch: Adaptive and Creative Projects for a Better City. Retrieved from: www.rtpi.org.uk/media/1642474/Placemaking-in-transitional-Christchurch-Rafeal.pdf

Mann, C. (2012). New riverside red zone, *The Press*, 23 March.

Mann, C. and Mathewson, N. (2012). New riverside red zone, *Stuff*, 23 March. Retrieved from: www.stuff.co.nz/the-press/news/christchurch-earthquake-2011/6624862/New-riverside-red-zone

Manyena, B., O'Brien, G., O'Keefe, P. and Rose, J. (2011). Disaster resilience: A bounce back or bounce forward ability? *Local Environment* 16 (5), 417–24.

Marcetic, B. (2017). New Zealand's neoliberal drift, *Jacobin*, 15 March. Retrieved from: www.jacobinmag.com/2017/03/new-zealand-neolioberalism-inequality-=welfare-state-tax-haven/

Marcuse, P. (2009). From critical urban theory to the right to the city, *City* 13 (2/3), 185–97.

Marques, B., McIntosh, J., Hatton, W. and Shanahan, D. (2019). Bicultural land-scapes and ecological restoration in the compact city: The case of Zealandia as a sustainable ecosanctuary. *Journal of Landscape Architecture* 14 (1), 44–53.

Marquis, F., Kim, J.J., Elwood, K.J. and Chang, S.E. (2017). Understanding post-earthquake decisions on multi-storey concrete buildings in Christchurch, New Zealand. *Bulletin of Earthquake Engineering* 15, 731–58.

Marsden, I. and Knox, G.A. (2008). Estuaries, harbours and inlets, in M. Winterbourn, G. Knox, C. Burrows and I. Marsden eds., *The Natural History of Canterbury*. Canterbury University Press, Christchurch, 735–70.

Marsh (2014). *Comparing Claims from Catastrophic Earthquakes*. Marsh & McLennan Companies, London.

Marshall, P. (2016). Redefining the garden city, in J. Harvey ed., *Christchurch – Five Years On, Architecture New Zealand*, March/April, 38–9.

Matan, A. and Newman, P. (2016). *People Cities. The Life and Legacy of Jan Gehl*. Island Press, Washington DC.

Matapopore (n.d.). Tūranga – The New Christchurch Central Library. Retrieved from: https://matapopore.co.nz/wp-content/uploads/2020/11/Final-Turanga.pdf

Matapopore Trust (2017). *Matapopore Cultural Design Strategy. The Ōtākaro Avon River*, Christchurch. Unpublished report.

Matthewman, S. and Byrd, H. (2020). A textbook case of how not to 'build back better'? Christchurch after the earthquakes (2010, 2011). Report presented at the World Urban Forum Abu Dhabi, February 2020.

Matthewman, S. and Uekusa, S. (2021). Theorising disaster communities, *Theory and Society* 50 (6), 965–84.

Matthews, P. (2012). Adrift in limbo land, *Stuff*, 4 February. Retrieved from: www.stuff.co.nz/the-press/news/christchurch-earthquake-2011/6365644/Adrift-in-limbo-land

Matthews, P. (2020). Munted: 'There was no dialogue, no transparency', *Stuff*, 2 September. Retrieved from: www.stuff.co.nz/the-press/news/122558553/munted-there-was-no-dialogue-no-transparency

McAloon, J. (2000a). The Christchurch elite, in J. Cookson and G. Dunstall eds., *Southern Capital: Christchurch. Towards a City Biography, 1850–2000*. Canterbury University Press, Christchurch, 193–221.

McAloon, J. (2000b). Radical Christchurch, in J. Cookson and G. Dunstall eds., *Southern Capital: Christchurch. Towards a City Biography, 1850–2000*. Canterbury University Press, Christchurch, 162–92.

McClure, J., Wills, C., Johnston, D. and Recker, C. (2011). How the 2010 Canterbury (Darfield) earthquake affected earthquake risk perception: Comparing citizens inside and outside the earthquake region, *Australasian Journal of Disaster and Trauma Studies* 2011 (2), 3–10.

McCrone, J. (2011). Chch CBD masterplan - a major risk? *Stuff*, 25 September. Retrieved from: www.stuff.co.nz/business/rebuilding-christchurch/5676846/Chch-CBD-masterplan-a-major-risk

McCrone, J. (2012a). Crunch time for central city blueprint, *Stuff*, 18 November. Retrieved from: www.stuff.co.nz/the-press/business/the-rebuild/7963020/Crunch-time-for-central-city-blueprint

McCrone, J. (2012b). Rebuilding to avoid drunken chaos, *Stuff*, 7 July. Retrieved from: www.stuff.co.nz/the-press/news/christchurch-earthquake-2011/7237990/Rebuilding-to-avoid-drunken-chaos

McCrone, J. (2014a). Christchurch's $600 million motorways, *Stuff*, 19 October. Retrievedfrom:www.stuff.co.nz/the-press/news/10632484/Christchurchs-600-million-motorways?rm=m

McCrone, J. (2014b), Christchurch rebuild: A city stalled, *Stuff*, 9 March. Retrieved from:www.stuff.co.nz/the-press/business/the-rebuild/9805314/Christchurch-rebuild-A-city-stalled

McCrone, J. (2015a). Christchurch retail precinct 'good as done', *Stuff*, 16 May. Retrievedfrom:www.stuff.co.nz/the-press/business/the-rebuild/68470623/christchurch-retail-precinct-good-as-done

McCrone, J. (2015b). The long wait for Christchurch's Metro Sports anchor project, *Stuff*, 26 June. Retrieved from: www.stuff.co.nz/the-press/business/the-rebuild/69718112/the-long-wait-for-christchurchs-metro-sports-anchor-project

McCrone, J. (2016a). Rolleston: Time to take it seriously, *Stuff*, 20 August. Retrieved from: www.stuff.co.nz/the-press/news/83246929/rolleston-time-to-take-it-seriously?rm=m

McCrone, J. (2016b). How the quakes changed the shopping geography of Christchurch, *Stuff*, 27 May. Retrieved from: www.stuff.co.nz/the-press/news/80398637/how-the-quakes-changed-the-shopping-geography-of-christchurch

McCrone, J. (2017). The future isn't going anywhere, so why did Christchurch rebuild the city of yesterday? *Stuff*, 26 August. Retrieved from: www.stuff.co.nz/the-press/news/95661149/the-future-isnt-going-anywhere-so-why-did-christchurch-rebuild-the-city-of-yesterday

McDonagh, J. (2014). Housing affordability in post-earthquake Christchurch. *Proceedings of the 20th Pacific Rim Real Estate Society Conference*, Lincoln, Canterbury, 19–22 January.

McDonagh, J. (2017). Shattered Dreams – Inner City Revitalisation, Gentrification and the Christchurch Earthquakes of 2010 and 2011. Unpublished PhD thesis, Lincoln University, Canterbury.

McDonagh, J. (2021). Personal communication, 23 August. John McDonagh was then honorary Associate Professor of Property, Lincoln University, Canterbury.

McDonald, L. (2012). Property owners leaving town, *Stuff*, 20 February. Retrieved from: www.stuff.co.nz/the-press/business/6440143/Property-owners-leaving-town

McDonald, L. (2013a). TC3 homes stigma in decline – industry, *Stuff*, 17 June. Retrievedfrom:www.stuff.co.nz/the-press/editors-picks/8802864/TC3-homes-stigma-in-decline-industry

McDonald, L. (2013b). Median house prices hit high, *Stuff*, 15 October. Retrieved from: www.stuff.co.nz/the-press/news/canterbury/9285625/Median-house-prices-hit-high

McDonald, L. (2013c). Who owns our city? *The Press*. 30 January, 1.

McDonald, L. (2013d). Key city investor pulls out, *Stuff*, 10 April. Retrieved from: www.stuff.co.nz/the-press/8492985/Key-city-investor-pulls-out

McDonald, L. (2015a). Developer slams council for sluggish city rebuild, *The Press*, 27 April, 8.

McDonald, L. (2015b). Turning point: Momentum in central city to swell after influx, *The Press*, 1 May, 1.

McDonald, L. (2016). Christchurch's $16.8m music venue The Piano ready for. August opening, *Stuff*, 8 June. Retrieved from: www.stuff.co.nz/business/80810200/newest-music-venue-filling-up

McDonald, L. (2017a). International flavours comes to High Street, *Stuff*. 28 May. Retrievedfrom:www.stuff.co.nz/the-press/business/92950946/international-flavours-comes-to-high-street

McDonald, L. (2017b). Revitalised Christchurch central city still seen as fragile, *Stuff*, 22 September. Retrieved from: www.stuff.co.nz/the-press/business/the-rebuild/ 96966272/revitalised-christchurch-central-city-still-seen-as-fragile

McDonald, L. (2017c). Council goes after central Christchurch's 'dirty 30' building owners who are 'kicking the city in the face', *Stuff*, 24 May. Retrieved from: www. stuff.co.nz/the-press/news/92863275/council-goes-after-central-christchurchs-dirty-30-building-owners-who-are-kicking-the-city-in-the-face

McDonald, L. (2018a). Project 8011 plans to repopulate central Christchurch, *Stuff*, 20 August. Retrieved from: www.stuff.co.nz/the-press/news/106380722/project-8011-plans-to-repopulate-central-christchurch

McDonald, L. (2018b). Sustainable apartment plan chosen for Crown-owned Christchurch site, *Stuff*, 21 December. Retrieved from: www.stuff.co.nz/business/109545173/ sustainable-apartment-plan-chosen-for-crownowned-christchurch-site

McDonald, L. (2018c). A matter of trust: Iwi voice in Christchurch rebuild, *Stuff*, 20 July. Retrieved from: www.stuff.co.nz/the-press/business/the-rebuild/103275206/a-matter-of-trust-iwi-voice-in-christchurch-rebuild

McDonald, L. (2019a). Central city living is slow to take off in Christchurch, *Stuff*, 26 October. Retrieved from: www.stuff.co.nz/business/116880859/central-city-living-is-slow-to-take-off-in-christchurch

McDonald, L. (2019b). Investors choosing apartments in Christchurch over pricey North Island homes, *Stuff*, 22 July. Retrieved from: www.stuff.co.nz/business/ property/114253407/investors-choosing-apartments-in-christchurch-over-pricey-north-island-homes

McDonald, L. (2019c). Finishing touches as Riverside Market prepares to open, *Stuff*, September 17. Retrieved from: www.stuff.co.nz/business/116086749/ finishing-touches-as-riverside-market-prepares-to-open

McDonald, L. (2019d). $500m for new Catholic cathedral, school, hotels and car parking in central Christchurch, *Stuff*, 7 December. Retrieved from: www.stuff.co.nz/ national/117961049/500m-for-new-catholic-cathedral-school-hotels-and-car-parking-in-central-christchurch

McDonald, L. (2020a). Group pulls out of Crown village plan, *The Press*, 2 December, 12.

McDonald, L. (2020b). Apartment sales pick up in central Christchurch's 'east frame', *Stuff*, 26 October. Retrieved from: www.stuff.co.nz/life-style/homed/real-estate/123147681/apartment-sales-pick-up-in-central-christchurchs-east-frame

McDonald, L. (2020c). New home boom in Chch worth $1 billion, *The Press*, 12 September, 1.

McDonald, L. (2020d). Christchurch nine years on: The reshaping of a city, *Stuff*, 22 February. Retrieved from: www.stuff.co.nz/national/119421238/christchurch-nine-years-on-the-reshaping-of-a-city

McDonald, L. (2020e). Emboldened geese invading city centre face population control, *Stuff*, 17 June. Retrieved from: www.stuff.co.nz/national/121676090/emboldened-geese-invading-city-centre-face-population-control

McDonald, L. (2021a). Central Christchurch site up for sale as 'sustainable village' anchor project finally abandoned, *Stuff*, 18 February. Retrieved from: www.stuff. co.nz/the-press/news/124273640/central-christchurch-site-up-for-sale-as-sustainable-village-anchor-project-finally-abandoned

McDonald, L. (2021b). Central Christchurch home building lags behind demand. *Stuff*, 1 March. Retrieved from: www.stuff.co.nz/life-style/homed/real-estate/ 124358871/central-christchurch-home-building-lags-behind-demand

McDonald, L. (2021c). More homes built but they are smaller, *The Press*. 13 March, 1.

McDonald, L. (2021d) Bid for 2000 new homes at Lincoln, *The Press* 20 March, 1.

McDonald, L. (2021e). Big players plan $90m precinct for key city site, *The Press*, 12 June, 1.

McDonald, L. (2022). CBD residents pass pre-quake numbers, *The Press* 14 February, 1.

McDowall, C. and Denee, T. (2019). *We Are Here. An Atlas of Aotearoa*. Massey University Press, Auckland.

McLean, I., Oughton, D., Ellis, S., Wakelin, B. and Rubin, C.B. (2012). Review of the Civil Defence Emergency Response to the 22 February 2011 Earthquake. Retrieved from: www.civildefence.govt.nz/resources/review-of-the-civil-defence-emergency-management-response-to-the-22-february-christchurch-earthquake/

{mds} law (2012). Newsletter: The Christchurch recovery plan. Retrieved from: www.mdslaw.co.nz/getattachment/e61bd861-41f8-41d7-aaea-364d9980af59/News-Publications-2012-The-Christchurch-Central-Recovery-Plan.aspx

Mead, T. (2017). Post-earthquake nightlife finally opening in Christchurch, *Newshub*, 11 January. Retrieved from: www.newshub.co.nz/home/new-zealand/2017/08/post-earthquake-nightlife-finally-opening-in-christchurch.html

Measures, R., Hicks, M., Shankar, U., Bind, J., Arnold, J. and Zeld, J. (2011). *Mapping Earthquake Induced Topographical Change and Liquefaction in the Avon-Heathcote Estuary*, Report No. U11/13. Environment Canterbury, Christchurch.

Meethan, K. (1996). Consuming (in) the civilized city, *Annals of Tourism Research* 23 (2), 322–40.

Meier, C. (2014). Christchurch rebuild ripoffs nail migrants, *Stuff*, 22 March. Retrieved from: www.stuff.co.nz/business/industries/9855596/Christchurch-rebuild-ripoffs-nail-migrants

Memon, P.A. (2003). Urban growth management in Christchurch, *New Zealand Geographer* 59 (1), 27–39.

Mental Health Foundation (n.d.). Five ways to wellbeing. Retrieved from: https://mentalhealth.org.nz/five-ways-to-wellbeing

Merkin, R. (2012). The Christchurch earthquakes insurance and reinsurance issues, *Canterbury Law Review* 18, 119–154.

Merritt, L. (2015). Personal communication, 11 March. Lauren Merritt was then Chief Executive Officer of the Ministry of Awesome.

Meurk, C.D. (2021). Think like a matai, in S. Goldsmith ed., *Tree Sense. Ways of Thinking about Trees*. Massey University Press, Auckland, 143–68.

Meurk, C. and Swaffield, S. (2000). A landscape ecological framework for indigenous regeneration in rural New Zealand-Aotearoa, *Landscape and Urban Planning* 50 (1–3), 129–44.

Meurk, C. and Swaffield, S. (2007). Cities as complex landscapes: Biodiversity opportunities, landscape configurations and design directions, *New Zealand Garden Journal* 10, 10–19.

Miles, S-A. (2016). *The Insurance Aftershock. The Christchurch Fiasco Post-earthquakes 2010–2016*, 2nd edn., Labyrinth Publishing, Christchurch.

Miller, S. (2012). Interest groups - Community and recreational groups. *Te Ara - the Encyclopedia of New Zealand*. Retrieved from: https://teara.govt.nz/en/interest-groups

Milligan, C. and Conradson, D. eds. (2006). *Landscapes of Voluntarism: New Spaces of Health, Welfare and Governance*. Policy Press, Bristol.

Ministry of Business, Innovation and Employment (2013). *Housing Pressures in Christchurch: A Summary of the Evidence*. Wellington, New Zealand. Retrieved from: www.mbie.govt.nz/dmsdocument/1088-housing-pressures-christchurch-2013-pdf

Ministry of Business, Innovation and Employment (2017). *Responses to the Canterbury Earthquakes Royal Commission Recommendations. Final Report*. Ministry of Business, Innovation and Employment, Wellington.

Ministry of Education (2021). *Rebuilding Christchurch Schools*. Retrieved from: www.education.govt.nz/our-work/changes-in-education/rebuilding-christchurch-schools/

Ministry for the Environment (1997). *The State of New Zealand's Environment 1997*. Ministry for the Environment/GP Publications, Wellington.

Ministry for the Environment (2008). *Urban Design Case Studies: Local Government*. Wellington. Retrieved from: https://environment.govt.nz/publications/urban-design-case-studies-local-government/

Mitchell, T. (2021). The Dirty 30 is down to the final six uildings. *Metronews*, 12 May. Retrieved from: https://metronews.co.nz/article/new-article-207

Monk, C.B., Van Ballegooy, S., Hughes, M. and Villeneuve, M. (2016). Liquefaction vulnerability increase at North New Brighton due to subsidence, sea level rise and reduction in thickness of the non-liquefying layer, *Bulletin of the New Zealand Society for Earthquake Engineering* 49 (4), 334–40.

Morgan, J., Begg, A., Beaven, S., Schluter, P., Jamieson, K., Johal, S., Johnston, D. and Sparrow, M. (2015). Monitoring wellbeing during recovery from the 2010–2011 Canterbury earthquakes: The CERA Wellbeing Survey, *International Journal of Disaster Risk Reduction* 14 (1), 96–103.

Morgenroth, J. (2017). Tree canopy cover in Christchurch New Zealand. Prepared for the Christchurch City Council. Retrieved from: https://ccc.govt.nz/assets/Documents/Environment/Trees/Tree-cover-in-Christchurch-final-report.pdf

Morgenroth, J. and Armstrong, T. (2012). The impact of significant earthquakes on Christchurch, New Zealand's urban forest, *Urban Forestry and Urban Greening* 11 (4), 383–9.

Moricz, Z., Wong, C. and Bond, S. (2012). *The Impacts of The Canterbury Earthquake on the Commercial Office Market*, CBRE and Lincoln University, Canterbury.

Morris, J.D.K. and Ruru, J. (2010). Giving voice to rivers: Legal personality as a vehicle for recognising indigenous peoples' relationships to water?, *Australian Indigenous Law Review* 14 (2), 49–62.

Mueller, S. (2017). Incommensurate values? Environment Canterbury and local democracy. *New Zealand Journal of Public and International Law* 15, 293–312.

Murphy, L. (2019). *Financiers and developers: Interviews concerning their interests, relationships and the residential development process*. Building Better Homes, Towns and Cities SRA1 – The Architecture of Decision-making Working Paper, School of Environment. University of Auckland, Auckland.

Murphy, L. (2020). Performing calculative practices: Residual valuation, the residential development process and affordable housing, *Housing Studies* 35 (9), 1501–17.

Murphy, S. (2015). Margaret Mahy ChCh playground opens, *Radio New Zealand*. Retrieved from: www.rnz.co.nz/news/regional/292727/margaret-mahy-chch-playground-opens

Mutch, C. (2017). Winners and losers: School closures in post-earthquake Canterbury and the dissolution of community, *Waikato Journal of Education* 22 (1), 73–95.

Mutch, C. (2018). The role of schools in helping communities copes with earthquake disasters: The case of the 2010–2011 New Zealand earthquakes, *Environmental Hazards* 17 (4), 331–51.

Naylor Love (2021). Our projects. New Regent Street, Christchurch. Retrieved from: www.naylorlove.co.nz/project/new-regent-street/

Neilson, M. (2021). Mental health: Minister Andrew Little 'extraordinarily' frustrated with slow spend on $1.9b package from 2019 Budget, *New Zealand Herald*, 22 June. Retrieved from: www.nzherald.co.nz/nz/mental-health-minister-andrew-little-extraordinarily-frustrated-with-slow-spend-on-19b-package-from-2019-budget/OG62VGSHJJHGRLF3SHPKM5HBJU/

Newell, J., Beaven, S. and Johnston, D.M. (2012). *Population Movements Following the 2010–2011 Canterbury Earthquakes: Summary of Research Workshops November 2011 and Current Evidence*. GNS Science Miscellaneous Series 44. GNS Science, Lower Hutt.

Newman-Storen, R. and Reynolds, R. (2013). Conversations over the gap, *Performance Research* 18 (3), 47–53.

New Zealand Government (2011). Govt outlines next steps for people of Canterbury. Retrieved from: www.beehive.govt.nz/release/govt-outlines-next-steps-people-canterbury

New Zealand Herald (2013). Four Christchurch schools to be replaced, 11 September. Retrieved from: www.nzherald.co.nz/nz/four-christchurch-schools-to-be-replaced/2GCEJTS36IBBRWAKU2K4TDVB7A/

New Zealand Police (2022). Christchurch earthquake. List of deceased. Retrieved from: www.police.govt.nz/news/major-events/previous-major-events/christchurch-earthquake/list-deceased

Nguyen, C. and Noy, I. (2017). Insuring earthquakes: How would the Californian and Japanese insurance programs have fared down under (After the 2011 New Zealand earthquake)? *SEF Working Paper 14/2017*, School of Economics and Finance, Victoria Business School, Wellington.

Nicholson, H. (2016). Cityscaping, in J. Harvey ed., *Christchurch – Five Years on, Architecture New Zealand* March/April, 47–8.

Nicholson, H. (2020). In G. Smyth, dir., *When a City Rises*, Frank Films. Retrieved from: www.nzonscreen.com/title/when-a-city-rises-2020

Nicholson, H. and Wykes, F. (2012). Share an Idea, *Urban Design* 123, 30–3.

Nielsen (2016). *Residential Red Zone Survey*. Nielsen, Auckland. Retrieved from: https://dpmc.govt.nz/sites/default/files/2017-03/cera-rrz-surveyreport-feb2016.pdf

Nissen, S., Carlton, S., Wong, J.H.K. and Johnson, S. (2021). 'Spontaneous' volunteers? Factors enabling the Student Volunteer Army mobilisation following the Canterbury earthquakes, 2010–2011, *International Journal of Disaster Risk Reduction* 53, 102008.

Norcliffe, J. and Preston, J. (2016). *Leaving the Red Zone: Poems from the Canterbury Earthquakes*. Clerestory Press, Christchurch.

Nowland-Foreman, G. (2016). Crushed or just bruised? Voluntary organisations - 25 years under the bear hug of government funding in Aotearoa New Zealand, *Third Sector Review* 22 (2), 53–69.

Noy, I., Kusuma, A. and Nguyen, C. (2017). Insuring disasters: A survey of the economics of insurance programs for earthquakes and droughts, *SEF Working Paper 11/2017*, School of Economics and Finance, Victoria Business School, Wellington.

NZLS (2010). Law Society comments on Canterbury Earthquake Response and Recovery Act, New Zealand Law Society, 30 September. Retrieved from: https://web.archive.org/web/20110720085134/http://www.lawsociety.org.nz/home/for_the_public/for_the_media/latest_news/news/september/law_society_comments_on_canterbury_earthquake_response_and_recovery_act

O'Callaghan, J. (2019). 'Things are not OK': Students at experimental school speak out, *Stuff*, 22 June. Retrieved from: www.stuff.co.nz/national/education/113205774/haeata-community-campus-students-begging-to-be-taught

O'Callaghan, J. (2021). February 2011: Earthquakes allow Christchurch's culture to change, *Stuff*, 22 February. Retrieved from: www.stuff.co.nz/pou-tiaki/300231617/february-2011-earthquakes-allow-christchurchs-culture-to-change

OECD (2020). *Innovative Citizen Participation and New Democratic Institutions. Catching the Deliberative Wave.* OECD Publishing, Paris.

OECD (2021). *Anticipatory Innovation Governance: What It Is, How It Works, and Why We Need It More Than Ever Before.* OECD Publishing, Paris.

Ombudsman (2018). Chief Ombudsman releases final opinion on East Lake Trust/Regenerate Christchurch case, 22 July. Retrieved from: www.ombudsman.parliament.nz/sites/default/files/2019-07/mediarelease-eastlakeregenerate.pdf

O'Neill, R. (2012). Compulsory acquisition could reconfigure Christchurch CBD, *Stuff*, 1 May. Retrieved from: www.stuff.co.nz/business/6831407/Compulsory-acquisition-could-reconfigure-Christchurch-CBD

OPES partners (2022). Canterbury property market, 27 January. Retrieved from: https://www.opespartners.co.nz/property-markets/canterbury

Orange, C. (1987). *The Treaty of Waitangi*. Allen and Unwin, Wellington.

Orchard, S. (2017). *Integrated Assessment Frameworks for Evaluating Large Scale River Corridor Restoration*. Report prepared for the Avon Ōtākaro Network, Christchurch.

Orchard, S., Hickford, M.J.H. and Schiel, D.R. (2018). Earthquake induced habitat migration in a riparian spawning fish has implications for conservation management, *Aquatic Conservation. Marine Freshwater Ecosystems* 28 (3), 702–12.

Otago Daily Times (2019). Government to pay up to $300 million for shoddy quake repairs, *Otago Daily Times*, 15 August. Retrieved from: www.odt.co.nz/news/national/govt-pay-300m-shoddy-quake-repairs

Otago Daily Times (2020). 10 years on: Homes still broken, insurance claims reopened, *Otago Daily Times*, 4 September. Retrieved from: www.odt.co.nz/star-news/star-christchurch/10-years-homes-still-broken-insurance-claims-reopened

Ōtākaro Limited (2021). Unparalleled development opportunities. Retrieved from: www.otakaroltd.co.nz/news-and-videos/unparalleled-development-opportunities/

Pain, R. (2019). Chronic urban trauma: The slow violence of housing dispossession, *Urban Studies* 56 (2), 385–400.

Parker, D. (2020). Disaster resilience – a challenged science, *Environmental Hazards* 19 (1), 1–9.

Parker, G. (2014). A new city through the arts?, in Bennett, B., Dann, J., Johnson, E. and Reynolds, R. eds., *Once in a Lifetime: City-building after Disaster in Christchurch*, Free Range Press, Christchurch, 338–41.

Parker, M. and Steenkamp, D. (2012). *The Economic Impact of the Canterbury Earthquakes*. Reserve Bank of New Zealand, Wellington.

Parliamentary Commissioner for the Environment (2015). *Preparing New Zealand for Rising Seas: Certainty and Uncertainty*. Office of the Parliamentary Commissioner for the Environment, Wellington.

Parsons, M. (2014). *Rubble to Resurrection: Churches Respond in the Canterbury Quakes*. DayStar, Auckland.

Paton, D., Selway, K.L., Mamula-Seadon, L. and GNS Science (2013). *Community Resilience in Christchurch: Adaptive Responses and Capacities During Earthquake Recovery*. GNS Science, Wellington.

Pauling, C., Awatere, S. and Rolleston, S. (2014). What has Ōtautahi revealed? Māori urban planning in post-earthquake Christchurch, in B. Bennett, J. Dann, E. Johnson and R. Reynolds eds., *Once in a Lifetime: City-building after Disaster in Christchurch*, Freerange Press, Christchurch, 452–64.

Pawson, E. (1999). Remaking places, in R. Le Heron, P. Forer, L. Murphy and M. Goldstone eds., *Explorations in Human Geography. Encountering Place*. Oxford University Press, Auckland, 345–66.

Pawson, E. (2000). Confronting nature, in J. Cookson and G. Dunstall eds., *Southern Capital Christchurch. Towards a City Biography 1850-2000*. Canterbury University Press, Christchurch, 60–84.

Pawson, E. (2011). Environmental hazards and natural disasters, *New Zealand Geographer* 67 (3), 143–47.

Pawson, E. (2022). Planning, governance and a city for the future? in S. Uekusa, S. Matthewman and B.C. Glavovic eds., *A Decade of Disaster Experiences in Ōtautahi Christchurch. Critical Disaster Studies Perspectives*. Palgrave Macmillan, Singapore, 317–34.

Pawson, E. and Christensen, A.A. (2014). Landscapes of the Anthropocene: From dominion to dependence? in J. Frawley and I. McCalman eds., *Rethinking Invasion Ecologies from the Environmental Humanities*. Routledge, London, 64–83.

Pawson, E., Kerr, R., Mein Smith, P. and Williams, C. (2019). *Ōtākaro Avon River Corridor. Governance Case Studies*. University of Canterbury, Christchurch.

Pawson, E. and Swaffield, S.R. (1998). Landscapes of leisure and tourism, in Perkins, H.C. and Cushman, G. eds., *Time Out? Leisure, Recreation and Tourism in Australia and New Zealand*, Longman Paul, Auckland, 254–70.

Peart, R. and Cox, B. (2019). *Governance of the Hauraki Gulf. A Review of Options*. Environmental Defence Society, Auckland.

Peat, N. (2018). *The Invading Sea. Coastal Hazards and Climate Change in Aotearoa New Zealand*. The Cuba Press, Wellington.

Pelling, M. (2010). Disaster politics: Tipping points for change in the adaption of socio-political regimes, *Progress in Human Geography* 34 (1), 21–37.

Percasky, M. (2020). Personal communication, 3 August. Mike Percasky was then a central city investor and developer involved in the Riverside Market and SALT initiative.

Percasky, M. and Shaw, G. (2020). Presentation to the Tuesday Club, Christchurch, July 21.

Perkins, H.C. and Rosin, C. (2018). Tourism, landscapes, and biological resources, in E. Pawson and the Biological Economies Team, *The New Biological Economy. How New Zealanders are Creating Value from the Land*, Auckland University Press, Auckland, 137–56.

Perkins, H.C. and Thorns, D.C. (2011). *Place, Identity and Everyday Life in a Globalizing World*, Macmillan, London.

Perrow, C. (1986). *Complex Organizations: A Critical Essay*. McGraw Hill, New York.

Pickles, K. (2016). *Christchurch Ruptures*. Bridget Williams Books, Wellington.

Place Leaders (2019). Place Leaders Asia Pacific, 2019 Place Leaders Awards. Retrieved from: www.placeleaders.com/post/2019-place-leaders-awards

Poontirakul, P., Brown, C., Noy, I., Seville, E. and Vargo, J. (2016). The role of commercial insurance in post-disaster recovery: Quantitative evidence from the 2011 earthquake. *Working Papers in Economics and Finance 01/2016*, School of Economics and Finance, Victoria Business School, Wellington.

Poontirakul, P., Brown, C., Seville, E., Vargo, J. and Noy, I. (2017). Insurance as a double-edged sword: Quantitative evidence from the 2011 Christchurch earthquake, *Geneva Papers on Risk and Insurance. Issues and Practice* 42 (4), 609–32.

Potongaroa, R., Wilkinson, S., Zare, M. and Steinfort, P. (2011). The management of portable toilets in the eastern suburbs of Christchurch after the February 22, 2011 earthquake, *Australasian Journal of Disaster and Studies* 2, 35–48.

Potter, S.H., Becker, J.S., Johnston, D.M. and Rossiter, K.P. (2015). An overview of the impacts of the 2010–2011 Canterbury earthquakes, *International Journal of Disaster Risk Reduction* 14 (1), 6–14.

Potts, A. and Gadenne, D. (2014). *Animals in Emergencies: Learning from the Christchurch Earthquakes.* Canterbury University Press, Christchurch.

The Press (2017). Editorial: Unleashing excitement in the red zone will make it Christchurch's field of dreams, August 30.

Pressman, J.L. and Wildavsky, A. (1984). *Implementation: How Great Expectations in Washington are Dashed in Oakland; Or, Why it's Amazing that Federal Programs Work at All, This Being a Saga of the Economic Development Administration as Told by Two Sympathetic Observers Who Seek to Build Morals on a Foundation* of Ruined Hopes (Vol. 708). University of California Press, Berkeley.

Preston, J. (2016). Report on the launch of Leaving the Red Zone. Retrieved from: https://joannapreston.com/2016/03/09/report-on-the-launch-of-leaving-the-red-zone/

Price, D. (2011). *Population and household trends in Christchurch post February 22 earthquake.* Population and Employment Effects of the Christchurch Earthquakes workshop. Lincoln University, Canterbury.

Project Lyttelton (2022). Welcome to Project Lyttelton. Retrieved from: www.projectlyttelton.org

Public Inquiry (2020). *Report of the Public Inquiry into the Earthquake Commission*, Earthquake Commission, Wellington. Retrieved from: https://dpmc.govt.nz/our-programmes/special-programmes/inquiry-eqc

Puig de la Bellacasa, M. (2017). *Matters of Care: Speculative Ethics in More Than Human Worlds.* University of Minnesota Press, Minneapolis.

Quake Outcasts (2016). Between Quake Outcasts and the Minister for Canterbury Earthquake Recovery, and the Chief Executive, Canterbury Earthquake Recovery Authority, CIV-2016-409-000050, NZHC 1959. Retrieved from: https://forms.justice.govt.nz/search/Documents/pdf/jdo/90/alfresco/service/api/node/content/workspace/SpacesStore/edc623c2-afe7-47ef-aa27-5a74a04a87b0/edc623c2-afe7-47ef-aa27-5a74a04a87b0.pdf

Quarantelli, E.L. (1999). *The Disaster Recovery Process. What We Know and Do Not Know from Research.* Disaster Research Center, University of Delaware, Newark, DE.

Quarantelli, E.L. and Dynes, R.R. (1977). Response to social crisis and disaster, *Annual Review of Sociology* 3 (1), 23–49.

Queenstown Airport Company (2021). Airport statistics. Retrieved from: www.queenstownairport.co.nz/corporate/airport-statistics

RNZ (2012). 13 schools to close, others to merge in Christchurch. Retrieved from: www.rnz.co.nz/news/canterbury-earthquake/115789/13-schools-to-close,-others-to-merge-in-christchurch

RNZ (2013). Inquiry into consultation on school restructuring progressing. Retrieved from: www.rnz.co.nz/news/national/214823/inquiry-into-consultation-on-school-restructuring-progressing

RNZ (2021). Hoiho Lane housing development opens in Christchurch. 9 July. Retrievedfrom:www.rnz.co.nz/news/national/446534/hoiho-lane-housing-development-opens-in-christchurch

Redzonecats (2021). Red Zone Cats. Retrieved from: https://redzonecats.org.nz

Regenerate Christchurch (2018a). *Whiti-Reia Cathedral Square. Our Long Term Vision*. Regenerate Christchurch, Christchurch.

Regenerate Christchurch (2018b). *Draft Ōtākaro Avon River Regeneration Plan*, Regenerate Christchurch, Christchurch.

Regenerate Christchurch (2019). *Ōtākaro Avon River Corridor Regeneration Plan*, Regenerate Christchurch, Christchurch.

Regenerate Christchurch (2021). Te Kōwhatawhata. About Us. Retrieved from: www.regeneratechristchurch.nz/about-us/

Resilient New Zealand (2015). *Contributing More: Improving the Role of Business in Recovery*. Retrieved from: https://iag.co.nz/content/dam/corporate-iag/iag-nz/nz/en/documents/corporate/business-role-in-recovery.pdf

Reynolds, R. (2014). Desire for the gap, in B. Bennett, J. Dann, E. Johnson and R. Reynolds eds., *Once in a Lifetime: City-building after Disaster in Christchurch*. Free Range Press, Christchurch, 167–76.

Reynolds, R. (2015). Looking backwards, thinking forwards. Retrieved from: www.gapfiller.org.nz/looking-backwards-thinking-forwards/

Rice, G. (2011). *All Fall Down. Christchurch's Lost Chimneys*. Canterbury University Press, Christchurch.

Roberts, M. (2006). From 'creative city' to 'no-go areas' - the expansion of the night-time economy in British town and city centres, *Cities 23* (5), 331–38.

Roe, J. and McCay, L. (2021). *Restorative Cities. Urban Design for Mental Health and Wellbeing*. Bloomsbury, London.

Rose, G. (1997). Situated knowledges: Positionality, reflexivities and other tactics, *Progress in Human Geography* 21 (3), 305–20.

Roughan, J. (2017). *John Key. Portrait of a Prime Minister*. rev. edn., Penguin Random House, Auckland.

Royal Commission (2012a). *Final Report*, Canterbury Earthquakes Royal Commission.Retrievedfrom:https://canterbury.royalcommission.govt.nz/Final-Report---Volumes-1-2-and-3

Royal Commission (2012b). *The Performance of Christchurch CBD buildings, Final Report, vol 2*. Canterbury Earthquakes Royal Commission. Retrieved from: https://canterbury.royalcommission.govt.nz/Final-Report-Volume-Two-Contents

Royal Commission (2012c). *Canterbury Television Building (CTV), Final Report, vol 6*. Canterbury Earthquakes Royal Commission. Retrieved from: https://canterbury.royalcommission.govt.nz/Final-Report-Volume-Six-Contents

Sachdeva, S. (2012). City looks ahead to much brighter future, *Stuff*, 2 January. www.stuff.co.nz/the-press/news/christchurch-earthquake-2011/6207203/City-looks-ahead-to-a-much-brighter-future

Sachdeva, S. and Carville, O. (2011). Christchurch rebuild plans 'rubbish', *Stuff*, 11 October. Retrieved from: www.stuff.co.nz/the-press/5763561/Christchurch-rebuild-plans-rubbish

Salmond, A. (2021). Three Waters and Te Tiriti, *Newsroom*, 22 December. Retrieved from: www.newsroom.co.nz/ideasroom/anne-salmond-three-waters-and-te-tiriti

Sanders, K. (2018). 'Beyond human ownership'? Property, power and legal personality for nature in Aotearoa New Zealand, *Journal of Environmental Law* 30 (2), 207–34.

Saunders, W.S.A. and Becker, J.S. (2015). A discussion of resilience and sustainability: Land use planning recovery from the Canterbury earthquake sequence, New Zealand, *International Journal of Disaster Risk Reduction*, 14, 73–81.

Schöllmann, A., Perkins, H.C. and Moore, K. (2000). Intersecting global and local influences in urban place promotion: The case of Christchurch, New Zealand, *Environment and Planning A* 32 (1), 55–76.

Schrader, B. (1997). Drains and plagues. Making Christchurch healthy, 1860s–1910s, in M. McKinnon ed., *New Zealand Historical Atlas*. David Bateman, Auckland, plate 55.

Scoop (2011a). Share an Idea wins international award, 24 November. Retrieved from: www.scoop.co.nz/stories/AK1111/S00632/share-an-idea-wins-international-award.htm

Scoop (2011b). Campaign for a memorial reserve covenant. Retrieved from: https://img.scoop.co.nz/media/pdfs/1108/CAMPAIGN.pdf

Scott, J.C. (1998). *Seeing Like a State. How Certain Schemes to Improve the Human Condition Have Failed*. Yale University Press, New Haven.

Seager, P. and Donnell, D. (2013). *Responders. The New Zealand Volunteer Response Teams Christchurch Earthquake Deployments*. Keswin Publishing, Christchurch.

Searle, W., McLeod, K. and Ellen-Eliza, N. (2015). *Vulnerable Temporary Migrant Workers: Canterbury Construction Industry*. Ministry of Business, Innovation and Employment, Wellington. Retrieved from: www.mbie.govt.nz/dmsdocument/2681-vulnerable-temporary-migrant-workers-canterbury-construction-pdf

Sessions, L.A. and Bullock, C. (2013). *Quake Dogs. Heart-Warming Stories of Christchurch Dogs*. Random House, Auckland.

Shaw, I. (2012). Towards an eventual geography, *Progress in Human Geography* 36 (5), 613–27.

Shaw, R. (2010). Neoliberal subjectivities and the development of the night-time economy in British cities, *Geography Compass* 4 (7), 893–903.

Sinclair, L. (2019). *Annual Review of the Greater Christchurch Regeneration Act 2016*, Department of Prime Minister and Cabinet, Christchurch.

Singer, M. (2020). Fendalton in Christchurch offers prestige properties, historic and new, *Mansion Global*, 30 May. Retrieved from: www.mansionglobal.com/articles/fendalton-in-christchurch-offers-prestige-properties-historic-and-new-215252

Škare, M., Soriano, D.R. and Porada-Rochoń, M. (2021). Impact of COVID-19 on the travel and tourism industry, *Technological Forecasting and Social Change* 163, 120469.

Small, J. and Meier, C. (2017). Labour announces $300m for Christchurch rebuild, *Stuff*, 27 August. Retrieved from: www.stuff.co.nz/the-press/business/the-rebuild/96197777/labour-announces-300m-for-christchurch-rebuild

Smith, A. (2017). The global game: Commodifying the playground at the end of the empire, *Interventions* 19 (7), 1056–67.

Smith, E. (2018). After seven years, who is scratching the itch in Christchurch's east? *Stuff*, 12 March. Retrieved from: www.stuff.co.nz/the-press/business/the-rebuild/101804585/after-seven-years-who-is-scratching-the-itch-in-christchurchs-east

Smith, J. (2014). Christchurch – a state of emergency, in B. Bennett, J. Dann, E. Johnson and R. Reynolds eds., *Once in a Lifetime. City-building After Disaster in Christchurch*. Freerange Press, Christchurch, 145–9.

Soens, T. (2020). Resilience in historical disaster studies: Pitfalls and opportunities, in M. Endress, L. Clemens and B. Rampp eds., *Strategies, Dispositions and Resources of Social Resilience*. Springer, Singapore, 253–75.

Solnit, R. (2009). *A Paradise Built in Hell. The Extraordinary Communities that Arise in Disaster*. Viking Penguin, New York.

Solomon, M. with Revington, M. (2021). *Mana Whakatipu. Ngāi Tahu Leader Mark Solomon on Leadership and Life*. Massey University Press, Auckland.

Spink, E. (2016). Visiting service set up for the elderly to curb loneliness, *Stuff*, 11 December. Retrieved from: www.stuff.co.nz/national/87383482/visiting-service-set-up-for-the-elderly-to-curb-loneliness

Spittlehouse, J.K., Joyce, P.R., Vierck, E., Schluter, P.J. and Pearson, J.F. (2014). Ongoing adverse mental health impact of the earthquake sequence in Canterbury, New Zealand, *Australian and New Zealand Journal of Psychiatry* 48 (8), 756–63.

Stallings, R.A. and Quarantelli, E.L. (1985). Emergent citizen groups and emergency, *Public Administration Review* 45, 93–100.

Statistics New Zealand (2013). *2013 Census data*. Retrieved from: https://stats.govt.nz/census/previous-censuses/2013-census/

Statistics NZ (2018). *Canterbury: The rebuild by the numbers*. Retrieved from: www.stats.govt.nz

Steeman, M. (2012). Business big guns join to shape retail future, *Stuff*, 27 October. Retrieved from: www.stuff.co.nz/the-press/news/7870830/Business-big-guns-join-to-shape-retail-future

Stengers, I. (2005). The cosmopolitical proposal, in B. Latour and P. Weibel eds., *Making Things Public: Atmospheres of Democracy*. MIT Press, Cambridge MA, 994–1003.

Stevenson, J. and Conradson, D. (2017). Organizational support networks and relational resilience after the 2010–2011 earthquakes in Canterbury, New Zealand, in Jones, E.C. and Faas, A.J. eds., *Social Network Analysis of Disaster Response, Recovery, and Adaptation*. Elsevier, Oxford, 161–76.

Stewart, A. (2013). Mahy-inspired playground wins top prize, *Stuff*, 23 May. Retrieved from: www.stuff.co.nz/the-press/editors-picks/8709026/Mahy-inspired-playground-wins-top-prize

Stoker, G. (1998). Governance as theory: Five propositions, *International Social Science Journal* 155, 17–28.

Strandh, V. and Eklund, N. (2018). Emergent groups in disaster research. Varieties of scientific observation over time and across studies of nine natural disasters, *Journal of Contingencies and Crisis Management* 26 (3), 329–37.

Stuff (2011). Quake know-how aid for Japan, *Stuff*, 24 April. Retrieved from: www.stuff.co.nz/the-press/news/4919917/Quake-know-how-aid-for-Japan

Stuff (2013). Editorial: Lessons from the school shake-up, *Stuff*, 20 February. Retrieved from: www.stuff.co.nz/dominion-post/comment/8324331/Editorial-Lessons-from-the-school-shake-up

Stuff (2015). Fletchers to build 191 inner-city homes for Christchurch, *Stuff*, 8 May. Retrieved from: www.stuff.co.nz/the-press/business/your-property/68403229/fletchers-to-build-191-inner-city-homes-for-christchurch

Stuff (2017). Editorial: Unleashing excitement in the red zone will make it Christchurch's field of dreams, *Stuff*, 30 August. Retrieved from: www.stuff.co.nz/the-press/business/the-rebuild/96310927/editorial-unleashing-excitement-in-the-red-zone-will-make-it-christchurchs-field-of-dreams

Stylianou, G. (2013). Tight-knit family behind tannery project, *Stuff*, 26 January. Retrieved from: www.stuff.co.nz/the-press/news/8226072/Tight-knit-family-behind-tannery-project

Stylianou, G. (2015). Report finds some migrant workers exploited, *Radio New Zealand*, 22 July. Retrieved from: www.rnz.co.nz/news/regional/279373/report-finds-some-migrant-workers-exploited

Stylianou, G. (2016). Regenerate Christchurch to investigate red zone water course, *Stuff*, 17 April. Retrieved from: www.stuff.co.nz/the-press/business/the-rebuild/79009732/regenerate-christchurch-to-investigate-red-zone-water-course

Sutch, W.B. (1966). *The Quest for Security in New Zealand, 1840–1966*, Oxford University Press, Wellington.

Swindell, D. and Rosentraub, M.S. (1998). Who benefits from the presence of professional sports teams? The implications for public funding of stadiums and arenas, *Public Administration Review* 58 (1), 11–20.

Tanner, T., Lewis, D., Wrathall, D., Bronen, R., Cradock-Henry, N., Huq, S., Lawless, C., Nawrotzki, R., Prasad, V., Rahman, M. and Alaniz, R. (2015). Livelihood resilience in the face of climate change, *Nature Climate Change* 5, 23–6.

Tatham, H. (2011). Social media key to 'army' success, *New Zealand Doctor* 15, 4 September.

Tau, T.M. ed., (2016). *Grand Narratives*. Canterbury Earthquake Recovery Authority, Christchurch.

Taylor, J.E., Chang, S.E., Elwood, K.J., Seville, E. and Brunsdon, D. (2012). *Learning from Christchurch: Technical Decisions and Societal Consequences in Post-Earthquake Recovery. Preliminary Research Findings*, August 2012. Resilient Organisations Programme, Christchurch.

Teller, C. (2008). Shopping streets versus shopping malls–determinants of agglomeration format attractiveness from the consumers' point of view, *The International Review of Retail, Distribution and Consumer Research* 18 (4), 381–403.

Tennant, M., O'Brien, M. and Sanders, J. (2008). *The History of the Non-profit Sector in New Zealand*. Office for the Community and Voluntary Sector, Wellington.

Te Putahi (2020). Te Putahi – about us. Retrieved from: https://teputahi.org.nz/about-us/

Thornley, L., Ball, J., Signal, L., Lawson-Te Aho, K. and Rawson, E. (2013). *Building Community Resilience: Learning from the Canterbury Earthquakes*. Final Report for Health Research Council and Canterbury Medical Research Foundation.

Thornley, L., Ball, J., Signal, L., Lawson-Te Aho, K. and Rawson, E. (2015). Building community resilience: Learning from the Canterbury earthquakes, *Kotuitui: New Zealand Journal of Social Sciences Online* 10 (1), 23–35.

The Press (2021). Commercial Property: Christchurch CBD office vacancy rate plunges to below pre-earthquake level, *The Press*, 22 September, 34.

The Star (2011). Earthquake 6.3 supplement. *APN New Zealand*, 22 February 2011.

Thompson, S., Espiner, S., Stewart, E., Shore, M. and Holland, P. (2015). *Banks Peninsula is the Stadium: Outdoor Recreation and Recovery in Post-earthquake Christchurch*, Lincoln University, Canterbury.

Thull, J-P. and Mersch, M. (2005). Accessibility and attractiveness – key features toward central city revitalisation – a case study of Christchurch, New Zealand, *Journal of the Eastern Asia Society for Transportation Studies* 6, 4066–81.

Tidball, K.G. (2014). Seeing the forest for the trees: Hybridity and social-ecological symbols, rituals and resilience in postdisaster contexts, *Ecology and Society* 19 (4), 25.

Till, K.E. (2012). Wounded cities: Memory-work and a place-based ethics of care, *Political Geography* 31 (1), 3–14.

Topp, C., Østergaard, S., Søndergaard, S. and Bech, P. (2015). The WHO-5 well-being index: A systematic review of the literature, *Psychotherapy and Psychosomatics* 84, 167–76.

Townsend, P. (2016). Personal communication 9, in M. Lesniak, Rebuilding/reimagining Post-disaster Christchurch – When Complexities Between Actors Have an Impact on Long-term Urban Development. Unpublished Masters Altervilles dissertation, Universite de Lyon, Lyon, 192–9.

Truebridge, N. (2017). Govt spent $1.5b acquiring Christchurch red zone land that's now worth just $21m, *Stuff*, 28 August. Retrieved from: https://www.stuff.co.nz/national/96144297/govt-spent-15-billion-acquiring-christchurch-red-zone-land-thats-now-worth-just-21m

Truebridge, N. (2018). Hagley Ave resident criticises Williams Corporation's planned Christchurch development for lack of car parking. *Stuff*, 9 February. Retrieved from: www.stuff.co.nz/business/better-business/101257861/hagley-ave-owner-criticises-williams-corporations-planned-christchurch-development

Uekusa, S., Matthewman, S. and Glavovic, B.C. eds. (2022). *A Decade of Disaster Experiences in Ōtautahi Christchurch. Critical Disaster Studies Perspectives.* Palgrave Macmillan, Singapore.

University of Washington (2020). *Ōtākaro Avon Landings. Tiaki Over Time: Ecocultural Regeneration + Climate Change Resilience for the Ōtākaro Avon Green Spine.* University of Washington, Department of Landscape Architecture, Seattle, Washington, Winter 2020.

Vallance, S. (2013). *Waimakariri District Council's Integrated, Community-based Recovery Framework.* Lincoln University, Canterbury.

Vallance, S. (2015). Disaster recovery as participation: Lessons from the Shaky Islands, *Natural Hazards* 75, 1287–1301.

Vallance, S. (2020). Personal communication, 12 March. Suzanne Vallance was then a board member of Greening the Rubble.

Vallance, S. and Carlton, S. (2015). First to respond, last to leave: Communities' roles and resilience across the '4Rs', *International Journal of Disaster Risk Reduction* 14, 27–36.

Vallance, S., Edwards, S., Conradson, D. and Karaminejad, Z. (2019). *Soft Infrastructure for Hard Times. Collaborative Planning for the (Re)building of Better Homes, Towns and Cities.* National Science Challenge: Building Better Homes, Towns and Cities, Porirua.

Vallance, S., Perkins, H.C. and Moore, K. (2005). The results of making a city more compact: Neighbours' interpretation of urban infill. *Environment and Planning B: Planning and Design* 32 (5), 715–33.

van Ballegooy, S., Malan, P., Lacrosse, V., Jacka, M.E., Cubrinovski, M., Bray, J.D., O'Rourke, T.D., Crawford, S.A. and Cowan, H. (2014). Assessment of liquefaction-induced land damage for residential Christchurch, *Earthquake Spectra* 30 (1), 31–55.

van Heugten, K. (2014). *Human Service Organizations in the Disaster Context.* Palgrave Macmillan, Basingstoke.

Veling, T. (2021). Vestiges. Retrieved from: www.timjveling.com/vestiges

Voxy (2011). New EQ community action group launched. Retrieved from: www.voxy.co.nz/national/new-eq-community-action-group-launched/5/103597

Voytenko, Y., McCormick, K., Evans, J. and Schliwa, G. (2016). Urban living labs for sustainability and low carbon cities in Europe: Towards a research agenda, *Journal of Cleaner Production* 133, 45–54.

Wade, P. (2018). World famous in New Zealand: Margaret Mahy Family Playground, Christchurch, *Stuff*, 7 October. Retrieved from: www.stuff.co.nz/travel/destinations/nz/canterbury/107502350/world-famous-in-new-zealand-margaret-mahy-family-playground-christchurch

Wagner, J. (1998). Canterbury Dialogues, *Our Environment. Christchurch City Council's Environmental Newsletter*, 14. Retrieved from: http://archived.ccc.govt.nz/OurEnvironment/14/

Waimakariri District Council (2017). Kaiapoi Regeneration areas/red zone notes, July 2017, Kaiapoi.

Waitākiri Eco-sanctuary (2022). Waitākiri sanctuary. He hiringa koiora ki Waitākiri. Retreived from: www.ecosanctuary.nz

Waka Kotahi New Zealand Transport Agency (2021a). Christchurch Motorways. Retrieved from: www.nzta.govt.nz/projects/christchurch-motorways/

Waka Kotahi New Zealand Transport Agency (2021b). Christchurch, Waimakariri and Selwyn Urban Cycleways Programme. Retrieved from: www.nzta.govt.nz/walking-cycling-and-public-transport/cycling/investing-in-cycling/urban-cycleways-programme/christchurch-urban-cycleways-programme/

Walker, B., de Vries, H.P. and Nilakant, V. (2017). Managing legitimacy: The Christchurch post-disaster reconstruction, *International Journal of Project Management* 35, 853–63.

Walsh, J. (2020). *Christchurch Architecture. A Walking Guide*. Massey University Press, Auckland.

Walton, S. (2020). Christchurch's 'Dirty 30' derelict sites still a work in progress, *Stuff*, 1 February. Retrieved from: www.stuff.co.nz/the-press/business/the-rebuild/1190 14561/christchurchs-dirty-30-derelict-sites-still-a-work-in-progress

Walton, S. (2021a). 'It's quite harsh out here': Data reveals clusters of social housing in Christchurch, *Stuff*, 8 May. Retrieved from: www.stuff.co.nz/the-press/news/125070869/its-quite-harsh-out-here-data-reveals-clusters-of-social-housing-in-christchurch

Walton, S. (2021b). Why hundreds of heritage buildings were demolished in Christchurch post-quake, *Stuff*, 22 February. Retrieved from: www.stuff.co.nz/the-press/business/the-rebuild/124157655/why-hundreds-of-heritage-buildings-were-demolished-in-christchurch-postquake

Walton, S. (2021c). City businesses split on parks, *The Press*, 25 March, 1.

Walton, S. (2021d). City may pitch for more stadium cash, *The Press*, 28 December, 2.

Warnock, A. (2014). Sam Johnson: leader of the Student Army, *Stuff*, 5 September. Retrieved from: www.stuff.co.nz/life-style/nz-life-leisure/61048171/sam-johnson-leader-of-the-student-army

Warren and Mahoney (2021). QEII Recreation and Sports Centre. Retrieved from: https://warrenandmahoney.com/portfolio/qeii-recreation-and-sports-centre

Weejes Sabella, E. (2015). Creative rebirth: Public art and community recovery in Christchurch. Retrieved from: https://hazards.colorado.edu/natural-hazards-observer/volume-xl-number-2

Weichselgartner, J. and Kelman, I. (2015). Geographies of resilience: Challenges and opportunities of a descriptive concept, *Progress in Human Geography* 39 (3), 249–67.

Welfare, R. (2020). Personal communication, 19 February. Rachael Welfare was then director of Life in Vacant Spaces.

Wellington City Council (2018). *Wellington Convention and Exhibition Centre. Business Case December 2018*. Wellington City Council, Wellington.

Wenger, C (2017). The oak or the reed: How resilience theories are translated into disaster management policies, *Ecology and Society* 22 (3), 18.

Wesener, A. (2015). Temporary urbanism and urban sustainability after a natural disaster: Transitional community-initiated open spaces in Christchurch, New Zealand, *Journal of Urbanism* 8 (4), 406–22.

Westbury, M. (2015). *Creating Cities*. Niche Press, Melbourne.

Whatmore, S. (2002). *Hybrid Geographies: Natures, Cultures, Spaces*. Sage, London.

Wilkinson, S. and Chang-Richards, A. (2016). *An Assessment of the Likely Changes to Construction Businesses in Christchurch Post Recovery*. BRANZ, Porirua. Retrieved from: www.branz.co.nz/pubs/research-reports/er7/

Williams Corporation (2021). Current Christchurch Developments. Retrieved from: www.williamscorporation.co.nz/current-developments-christchurch/

Wilson, G.A. (2012). *Community Resilience and Environmental Transitions*. Routledge, Abingdon.

Wilson, G.A. (2013). Community resilience, social memory and the post-2010 Christchurch (New Zealand) earthquakes, *Area* 45 (2), 207–15.

Wilson, G.A. (2014). Community resilience: Path dependency, lock-in effects and transitional ruptures, *Journal of Environmental Planning and Management* 57 (1), 1–26.

Wilson, J. (1984). *Lost Christchurch*. Te Waihora Press, Lincoln.

Wilson, J. (1989). *Christchurch, Swamp to City. A Short History of the Christchurch Drainage Board 1875–1989*. Te Waihora Press, Lincoln.

Winn, C. (2015). Personal communication, 2 March. Coralie Winn was co-founder of Gap Filler.

Wisner, B. (2020). Vulnerability, in A. Kobayashi ed., *International Encyclopedia of Human Geography*, 2nd edn., Elsevier, Amsterdam, 14, 197–205.

Wisner, B., Blaikie, P., Cannon, T. and Davis, I. (2004). *At Risk. Natural Hazards, People's Vulnerability and Disasters*, 2nd edn., Routledge, New York.

Wisner, B. and Luce, H.R. (1993). Disaster vulnerability: Scale, power and daily life, *GeoJournal* 30 (2), 127–40.

Witsey, M. (2016). Student Volunteer Army founder Sam Johnson to expand service, *Stuff*, 31 May. Retrieved from: www.stuff.co.nz/southland-times/news/80584789/student-volunteer-army-founder-sam-johnson-to-expand-service

Wolch, J. (1990). *The Shadow State: Government and Voluntary Sector in Transition*. Foundation Center, New York.

Wood, A. (2011). Go-ahead for Lyttelton Port project, *Stuff*, 25 May. Retrieved from: www.stuff.co.nz/business/5052556/Go-ahead-for-Lyttelton-Port-project

Wood, A. (2012). Unit to attract private sector into rebuild, *Stuff*, 31 July. Retrieved from: www.stuff.co.nz/business/7380309/Unit-to-attract-private-sector-into-rebuild

Wood, A. (2015). Fletcher Residential wins $800m Christchurch housing project, *Stuff*, 2 July. Retrieved from: www.stuff.co.nz/business/industries/69904053/fletcher-residential-wins-800m-christchurch-housing-project

Wood, A. and Meier, C. (2015). Fletcher and Government partner on $800 million housing project, *Stuff*, 2 July. Retrieved from: www.stuff.co.nz/business/69915244/fletcher-and-government-partner-on-800-million-housing-project

Wood, A., Noy, I. and Parker, M. (2016). The Canterbury rebuild five years on from the Christchurch earthquake, *Reserve Bank of New Zealand Te Pūtea Matua Bulletin* 79 (3), 1–16.

Woodford, A. (2019). Black-billed gull colony in Central Christchurch. Retrieved from: www.annettewoodford.wordpress.com

Woods, M. (2017). Minister announces new plan for metro sports facility. Ministerial Announcement 27 November. Retrieved from: www.beehive.govt.nz/release/minister-announces-new-plan-metro-sports-facility

Woods Bagot (2021). River recovery. Retrieved from: www.woodsbagot.com/projects/christchurch-convention-centre/

Wotif (2021). Hotels near Margaret Mahy Playground. Retrieved from: www.wotif.co.nz/Margaret-Mahy-Playground-Hotels.0-1553248635975795014-0.Travel-Guide-Filter-Hotels

Wright, M. (2001). *Quake: Hawkes Bay 1931*. Reed, Auckland.

Wright, M. (2012). Global travel guide hails Chch as top destination, *Stuff*, 22 October. Retrieved from: www.stuff.co.nz/the-press/news/7845845/Global-travel-guide-hails-Chch-as-top-destination

Xayasenh, P. (2015). Post-quake recreational activities in the Avon River and the Avon-Heathcote Estuary. Unpublished Master of Science Thesis, University of Canterbury, Christchurch.

Young, C. (2019). Fletcher Living homes in Central Christchurch struggle to find buyers, *Radio New Zealand*, 10 June. Retrieved from: www.rnz.co.nz/national/programmes/checkpoint/audio/2018698951/fletcher-living-homes-in-central-christchurch-struggle-to-find-buyers

Yusoff, K. (2013). The geoengine: Geoengineering and the geopolitics of planetary modification, *Environment and Planning A: Economy and Space* 45 (12), 2799–808.

Zaki, A. (2021), Christchurch CBD dwelling goal 'unrealistic', councillor says. *Radio New Zealand*, 14 May. Retrieved from: www.rnz.co.nz/news/national/442562/christchurch-cbd-dwelling-goal-unrealistic-councillor-says

Zukin, S. (1998). Urban lifestyles: Diversity and standardisation in spaces of consumption, *Urban Studies* 35 (5–6), 825–39.

Index

Pages in *italics* refer to figures; pages in **bold** refer to tables.

Printed in the United States
by Baker & Taylor Publisher Services

Printed in the United States
by Baker & Taylor Publisher Services